Metayage, Capitalism and Peasant Development in St Lucia 1840-1957

Peter Adrien

Consortium Graduate School of Social Sciences
UNIVERSITY OF THE WEST INDIES
MONA JAMAICA

Consortium Graduate School of Social Sciences
The University of the West Indies
Mona Campus Kingston 7 Jamaica
ISBN 976-41-0089-9
ISSN 0799-0057

in collaboration with Canoe Press University of the West Indies
Barbados · Jamaica · Trinidad and Tobago

This title in the New Generation Series has been published with financial assistance from the Caribbean Development Bank

Printed in Canada

00 99 98 97 96 5 4 3 2 1

CATALOGUING IN PUBLICATION DATA

Adrien, Peter
Metayage, capitalism and peasant development in St Lucia, 1840-1957 / Peter Adrien.

p. cm. – (New generation series ; 4)
Includes bibliographical references.
ISBN 976-41-0089-9
ISSN 0799-0057
1. Sharecropping – Saint Lucia – History.
2. Peasantry – Saint Lucia – History.
3. Saint Lucia – History. 4. Agriculture – Economic aspects – Saint Lucia. I. Title.
HD1478.S17A38 1996 306.365 dc–20

Book design by Orville Bloise
Set in 10.5 x 13 Palatino x 27

Metayage, Capitalism and Peasant Development in St Lucia 1840-1957

CONSORTIUM GRADUATE SCHOOL OF SOCIAL SCIENCES
NEW GENERATION SERIES
Norman Girvan, Series Editor
Bill Riviere, Associate Series Editor

This series brings to the attention of academic readers and the general public the work of young Caribbean scholars on a broad range of topics related to Caribbean development, orginally undertaken for research degrees at the Consortium Graduate School.

1 *Idology and Caribbean Integration*
Ian Boxill

2 *Technological Behaviour under Structural Adjustment: a Case Study*
Girjanauth Boodraj

3 *Gender Segregation in the Barbadian Labour Market 1946 and 1980*
Roslyn Lynch

4 *Metayage, Capitalism and Peasant Development in St Lucia 1840-1957*
Peter Adrien

CONTENTS

LIST OF TABLES

Dedicated to my wife, Michella Adrien,
my children, Miriam and Moses,
and the resilient peasants who
orchestrated the silent revolution
in St Lucia in the postemancipation period

FOREWORD

Peter Adrien's *Capitalism, Metayage and Peasant Development* is a study of the rise and fall of *metayage* in Dennery, a rural community in St Lucia during the period from Emancipation to the middle of the twentieth century. Metayage was a form of landholding based on sharecropping that was forced upon former slave owners as a result of a scarcity of labour caused by the transition to legal freedom. Relying on primary and secondary archival and oral sources, the author explores the origins and establishment of *metayage,* its consolidation and expansion while the plantocracy declined, its impact on economic and social life, and its demise and the concomitant rise of a peasantry.

The author challenges the classical and neoclassical view of sharecropping as a temporary and convenient form of agricultural production as well as the Marxian proposition that it represented a capitalist experiment in which the labourers were victims. By contrast, he strongly argues that *metayage* was a critical form of agricultural production for a hundred years following the emancipation of slaves and that while *metayers* were indeed exploited they were in a sense victors, for sharecropping provided a foundation for accumulation that promoted the development of a native agrarian 'capitalist class' and altered the character of social, economic and political life in Dennery and in St Lucia. The author's analysis of the impact of the *metayage* "form of production" on the "social and structural transformation" of society is supported on the notion that "the dominant form of production in society provides the motor for social change". The study also considers some policy implications for the present.

Publication of the study is timely. It makes a useful contribution to St Lucian and Caribbean historiography by updating the experience of *metayage* in the Caribbean to the twentieth century rather than treating it, in the manner of earlier studies, as an essentially postemancipation survival strategy by ex-slaves. It portrays *metayage* as a form of land tenure alternative to the plantation, co-existing with the plantation, in conflict yet mutually dependent upon it and, assisted by exogenous circumstances, finally ousting the plantation as the dominant form of agricultural production. In that sense the study is a story of economic and social resistance by black former slaves against colonial rule. Further, in light of the demise of socialism and the need for appropriate paradigms, the author's attempt to use Marxist methodology creatively rather than slavishly is a source of food for thought.

BILL RIVIERE
APRIL 1996

PREFACE

This work traces the development of an agrarian 'capitalist' class in St Lucia between 1840 and 1957.

Based on the assumption that the dominant form of production in society provides the motor for social change, the study posits the view that the *metayage* system (practised extensively in export agriculture between 1840 and 1957) set in train mechanisms for the changes which ensued.

The central argument of the study is that the *metayer* class was the 'feeder class' from which the 'capitalist' peasant classes which presently dominate the export agricultural sector evolved. It shows that the character of the accumulation process under *metayage* (among other factors) led to the emergence of an agrarian 'capitalist' class from the labouring population which altered the character of the social structure in the colony.

The data analysed suggest an increasing differentiation among the labouring class and the emergence of a large and influential differentiated free peasantry under *metayage*.

The study considers some theoretical and policy issues in the final chapter.

ACKNOWLEDGEMENTS

A number of persons contributed directly and indirectly to the development of the thoughts expressed in this book, but the author assumes sole responsibility for the issues raised. For her exemplary level of tolerance and incisive criticism as a supervisor, Dr Elsie Le Franc deserves my profound gratitude.

I am indebted to all core and associate faculty staff and students of the Consortium Graduate School of Social Sciences for their academic support, particularly Professor Norman Girvan.

Special mention must be made of Dr Bill Riviere, for his unselfish assistance and insightful critique of some of the chapters and Dr Hilary Beckles at the University of the West Indies, Cave Hill campus, for his comments on the chapters of analysis.

To the many public servants, librarians, archivists and others who facilitated my research in Barbados, Grenada, Jamaica, St Lucia, and Trinidad and Tobago, I express my sincere appreciation. Heartfelt appreciation also goes to Mrs Margarette Pearce at the University of the West Indies Main Library and Messrs Gregor Williams and Robert Deveaux of the St Lucia National Archives.

PETER ADRIEN
NOVEMBER 1995

CHAPTER ONE

Introduction

The Problem

This work argues that the dominant form of production in any specific historical period is the vehicle for social and structural change in any society [Abalkin 1983]. This motor of social change in St Lucia between 1840 and 1957 was the *metayage* system, a French form of sharecropping introduced during the French occupation. *Metayage*, a form of agricultural production in the immediate postemancipation era, set in train mechanisms that altered the physical, social, and economic structure of St Lucia. It is noteworthy that the effects of *metayage* on the process of social transformation in St Lucia have never been acknowledged by planners, in spite of the fact that the practice featured significantly in export agricultural production in the colony from 1840 to 1957. There seem to be three major reasons for the omission or neglect:

1. Studies on *metayage* (except for the work of Woodville Marshall [1963]) have failed to recognize *metayers* as a significant force in social change in general, and in rural development in particular, even in former colonies like St Lucia, where *metayage* was practised extensively. The tendency is to portray the system merely as a mid nineteenth century phenomenon, and to treat *metayers* as victims rather than victors [Hall 1959; Marshall 1963; Dookhan 1977; Green 1981; Louis 1981].
2. Classical and neoclassical economists continue to treat sharecropping as an economically inefficient form of production and, as a result, place

very little analytical emphasis on it or any of its multiple forms that have survived [Smith 1976; Marshall 1961].

3. In the main, most Marxist theorists place very little emphasis on the development of sharecropping. This form of production is perceived as a stubborn remnant of feudalism which will inevitably be destroyed with the development of capitalism in agriculture [Marx 1967, 1986; Pearce 1983]. However, Lenin discussed the development of small and middle peasant capitalists under sharecropping in prerevolutionary Russia [Lenin 1964: 159-286].

The following analysis argues that the contradictions that emanated from the relations of production and the process of accumulation within *metayage*, a quasi feudal form of production inserted within a dominant underdeveloped and weak capitalist mode of production [Marshall 1963] in the period 1840 to 1957, set in train the mechanisms that directed the character of the changes in the constitution of the social structure.

The growth of the *metayage* system was conditioned by the fluctuations in demand and supply, price fluctuations on the international commodity market, and the changes in the responses of capital to the resulting cycles of crises and depressions in the national and global economies. The practice expanded with each period of downswing, and contracted with each period of upswing. In each cycle the character of production and accumulation was conditioned by the demand for, and supply of, labour and capital. In the first phase, the share of the *metayers* in the accumulation process significantly altered the nature of property relations in the rural communities in the colony. The second phase was much less beneficial to the *metayers*. There were two reasons for this. First, the planters preferred to cultivate cocoa on share arrangement because it allowed for greater exploitation of labour and, secondly, the dislocations of World Wars I and II caused serious shortfalls in world production of sugar, and thereby reduced the level of acreage under *metayage* cultivation in the sugar industry.

The study poses the hypothesis that *metayage* provided the mechanisms that promoted the emergence of a 'capitalist' class in agriculture in St Lucia between 1840 and 1957. The term 'capitalist' refers neither to the industrial capitalist discussed by Marx nor the bourgeois capitalist peasant discussed by Lenin. The concept 'capitalist' is used primarily to refer to the involvement and integration of the peasant classes into the capitalist system and their utilization of capitalist logics of accumulation under capitalist and precapitalist arrangements.

Our main task is to find out the extent to which the *metayage* system as a form of production altered the following:

1. The organization of production
2. The nature of the relations of production
3. The character of the accumulation process
4. The character of labour and capital to the extent that it set in train mechanisms which accelerated the decline of the planter class and the rise of a 'capitalist' class

Methodology

Primary and secondary sources of statistical data are emphasized, since measurable data are required to test the working hypotheses. Insofar as the task requires the analysis of the problem over a specific period of time, oral history and archival research are also utilized to complement the quantitative techniques of data collection.

Inasmuch as the study analyses developments during the colonial era, the data were collected mainly from colonial documents. Secondary data were extracted from Colonial Office Reports, the *Blue Books of Statistics, Royal Commission Reports,* Intercolonial Dispatches, Stipendiary Magistrates Reports, and so on.

To place the phenomenon under study in a broader frame of reference, information from the archives in Grenada (at the residence of the Governor General) and Tobago, are used for purposes of comparison. Grenada, Tobago and St Lucia were part of the same geopolitical region, shared the same colonial administration, and practised the *metayage* system successfully in the nineteenth century [Marshall 1963; Dookhan 1977].

Oral data are utilized to validate the information contained in official reports, as well as to tap the many non-measurable dynamics, which official reports always fail to capture. The historian and the social scientist have the same task (it is lamentable that some historians think otherwise), that is, to grapple with social reality. Therefore, it is expected that "both historians and social scientists are interested in regularities, tendencies or receptive elements in social behaviour" [Shafer 1971: 6].

Much of the data came from interviews. A sample of sixteen landed families, labourers, and other resource persons was interviewed. The respondents comprised sugar workers, landlords and *metayers* presently engaged in *metayage*, persons whose families were involved in *metayage* cultivation in the period under study, as well as other persons who were actively involved in electoral politics, education and commerce in the district in the same period. In addition to the residents of the district, two resource persons who resided outside of the district, but worked in the area (a head teacher and a tax officer) during the sugar era, were also interviewed.

This work assumes a particular structural arrangement. Chapter 2 offers a short profile of St Lucia so as to provide a reference point for the broader socioeconomic analysis.

Chapter 3 begins by attempting an analysis of the theoretical debate surrounding sharecropping, and then proposes a theoretical framework within which *metayage* in St Lucia between 1840 and 1957 can be adequately examined.

The three chapters of analysis (chapters 4, 5, and 6), which correspond with the three phases of the *metayage* system, follow a specific pattern. Each chapter begins with an examination of the historical context, assesses the dynamics of *metayage,* and ends with a summary of the social and structural concerns discussed. Chapter 4 examines the development problem in highland and lowland Dennery, highlighting the current disparity in the growth patterns of the two areas. This appraisal is done within the broader context of the wider national economy.

Chapters 4, 5, and 6 deal with the central focus of this work. The available data are advanced to explain the nature, course and cause of the reconstitution of social structure in the period 1840 to 1957. The analysis is done in three stages. They are marked by specific economic responses to international commodity market shocks, the impact of the shocks on the character of *metayage,* and the consequent effect on the society. Chapter 4 covers the period 1840 to 1897, and is referred to as the period of the establishment of *metayage*. Chapter 5 examines the second phase of *metayage* expansion, 1897 to 1929. Chapter 6 evaluates the final phase of extensive *metayage* practice, 1929 to 1957.

Chapter 7 concludes the study. It analyses the cogency of the hypotheses, provides a summary of the central problem, outlines the course and cause of the reconstitution of the social structure in St Lucia for the period under discussion, and makes prescriptions specifically for more equitable and sustained development of the country. Attempts are made to recommend management prescriptions for planning and implementation of development programmes for the districts in St Lucia where *metayers* played a significant role in the process of social change.

CHAPTER TWO

St Lucia: A Profile

St Lucia is approximately 238 square miles in size. Located in the group of islands of the English-speaking Caribbean, it is a mountainous island, one of a chain of volcanic islands in the Eastern Caribbean. Its nearest neighbours are Martinique in the north-east and St Vincent in the south-east. The main mountain ranges run north and south for almost the entire length of the island.

There are numerous plains and valleys irrigated and fertilized by a number of rivers and streams; the Dennery river is one of the principal rivers of the island. The three principal valleys, Roseau and Cul-de-Sac in the west and Mabouya (in Dennery) in the east, formed from the erosion of volcanic materials, are used mainly for agricultural production.

St Lucia was first colonized by France and changed hands fourteen times before it was ceded in 1815 to the British during the Napoleonic War. However, the French custom continues to influence social and cultural practices in the society. The French influence is expressed in the language spoken *(Creole)*, the land tenure system, and established non-formal institutions for labour sharing *(coud de main)*, for capital mobilization *(sou sou)* and agricultural production *(metayage)*.

The population of St Lucia is relatively small, but has been increasing steadily over the last few decades. The population grew by 20 percent to 135,685 in 1991, from 115,252 persons in 1980. Since 1970, the population has increased at a rate of approximately 1,500 persons per year. The greater proportion of the population is made up of persons less than 50 years of age. From 1911 to 1986 the population change has been positive. Despite an infant mortality rate of 20.1, the birth rate was 26.1 in 1990.

Life expectancy is 69 years for males and 74 years for females, and compares favourably with most developed countries. The health sector includes 25 health centres and 5 hospitals. However, they are inadequate, as they cannot supply the increasing demands of the expanding population.

The adult literacy rate (15 to 65 age group) stands at 46 percent [*St Lucia Human Settlements* 1988: 1]. Although English is the official language as well as the language of instruction, French patois is the dominant language of communication, particularly in the rural communities.

The economy is small, open, dependent, and dominantly agrarian. However, tourism, an expanding sector, contributes significantly to the gross national product. In spite of the slowdown in economic activity caused by the global recession in the 1970s, all the major sectors of the economy registered significant growth in the second half of the period. Economic activity during that period was largely driven by the construction of the Hess Oil transshipment terminal. Although economic growth was adversely affected in 1980 as a result of the passage of Hurricane Allen, the boom in the banana industry from the mid 1980s to the early 1990s produced a sustained period of economic growth. The economy grew at an average growth rate of 5 percent during that period.

The island lacks a well developed road network, due to the fact that it is very rugged and mountainous. Road maintenance consumes a significant portion of government revenue, as the island is disadvantaged by a volcanic soil that is easily eroded by running water. Notwithstanding the constraints, there are four types of roads: main, secondary, tertiary, and unpaved. Those in the urban centres are of the main and secondary types. The East Coast highway connects the northern urban centre (Castries, the capital city) to the southern urban centre (Vieux-Fort) 42 miles away. The West Coast highway which will replace the narrow colonial road connecting the western urban centre (Soufrière) with the north and the south, is almost completed.

Air and seaports have been further developed in order to accomodate the expanding tourist industry and the banana trade. Banana shipment is done via the Vieux-Fort port and the Castries harbour, while Vigie Airport in Castries and the Hewanorra International Airport in Vieux-Fort provide modern facilities for small, medium and large aircraft.

As is the case with many other Caribbean economies, the traditionally small manufacturing sector is characterized by a very high import content in its product mix [Davies 1987]. The manufacturing sector, in recent times, suffered adversely from the loss of major regional markets as a result of the imposition of trade restrictions. Among the commodities produced, textile goods, paper products, and rum are the most important.

The financial sector is very dependent on the external financial community. Most commercial banks and insurance companies are branches of parent companies incorporated in Europe, North America, and the other Caribbean countries. The Eastern Caribbean dollar (EC$), issued by the Eastern Caribbean Central Bank (ECCB) and pegged to the US dollar at the rate of EC$ 2.70 to US$ 1.00, is the official currency. St Lucia has no central bank of its own, but is a member of the ECCB and benefits from the multistate central banking arrangement that was ratified by the governments of the Eastern Caribbean.

Historically, the agricultural sector has been the major contributor to the gross national product, but the pattern of acquisition and ownership of cultivable land has always militated against its development. The system of land tenure has features that are subject to customary symbolic valuation and formal legal definitions. In St Lucia, land functions not simply as a legally defined economic good, but also as a medium for the expression and sustenance of social relations.

The problems of inheritance, such as conflict over the generational distance over which land should be inherited and disagreements over the use of family land, act as impediments to agricultural development in the rural areas. The rights to harvest without obligation to investment, the inability to use the land as collateral for securing credit, and the frequently, fragmented relationships among kinsmen militate against productivity and proper practice of conservation and protection.

The land system in St Lucia is a legacy of the Code Napoleon, established after the French Revolution and the Napoleonic wars started in 1789 [Brook 1976: 30]. "Before the Civil Code came into force the primary source of our civil law (which includes our land law) was mainly the Coutume de Paris" [Cenac 1988: 5]. The Napoleonic Code grants a surviving spouse and children the right to a parcel of land upon the death of the owner. A large family, therefore, may share a single parcel of five acres [Romalis 1968: 98-101].

Family land is the most dominant customary land tenure type, apart from the five other formal legal and customary definitions of tenure types in St Lucia. Generally, people occupy land by legal ownership, leasehold, permission, squatting and as sharecroppers. What distinguishes family land from the other tenure types is the fact that it is created or perpetuated with intestate succession, and is the collective inheritance of possession, use, and usufruct rights in land by a group of persons who are related by blood.

In St Lucia, land as a factor of production that is crucial to rural and urban development is a scarce resource. It is estimated that only about 9 percent or 14,000 acres of the land area is arable, of which 86 percent or 12,000 acres is under banana cultivation. Of the 12,000 acres under cultivation, more than 10,000 holdings are less than 5 acres in size-holdings occupied by small farmers. Small farmers, the major contributors in the agricultural sector, control only 1 percent of the farm land area, while middle and large farmers control 32 percent and 67 percent, respectively [Latin American Bureau 1987: 60-67].

The agricultural sector has been the most dominant of the productive sectors and contributes the most to gross domestic product (GDP) and foreign exchange generation. Between 1985 and 1988 the economy grew at an average rate of 5.5 percent. In 1987 GDP at constant prices amounted to EC$ 237.7 million, of which agriculture and tourism contributed 37.4 percent and 18.0 percent respectively. The trade deficit was at a high of EC$283 million, but the inflation rate was at a low of 0.82 percent.

The banana industry is a significant contributing factor to the economic and social development of the island. The character of the industry, that is to say, the fact that the small and medium peasant farmers are the major producers and accumulators in the industry, allows for more equitable distribution of income among the rural population relative to that under sugar. The medium farmers (those who cultivate holdings between 6 and 50 acres in size), who enjoy easy access to the labour and money market, benefit most from the industry, although "small peasants owning less than five acres of land own 82.1 percent of all farms" [Latin American Bureau 1987: 67].

In spite of the gains from the banana industry and its multiplier effect on the network of production, distribution, and consumption, the economy, like that of most developing countries, continues to be characterized by dualism, a skewed resource base and lack of forward and backward linkages [Beckford 1975]. The lack of linkages between the sectors of the economy, particularly between agriculture and tourism, implies fewer employment opportunities for the many thousands of school leavers each year as well as a spiraling food import bill. The literature reveals that "the economy cannot provide sufficient jobs for those able to work". Although the unemployment rate is estimated at 20 percent to 35 percent of the labour force, it is probably nearer 35 percent or 40 percent, if the large number of people who are underemployed are included [Latin American Bureau 1987: 60].

CHAPTER THREE

Metayage: Towards a Theoretical Framework

This chapter attempts to appraise the longstanding economic and historical discourse on sharecropping in general, and *metayage* in particular, with the view of suggesting an alternative perspective to the analysis of *metayage* in St Lucia.

Much of the discussion on sharecropping arose initially out of the concern for economic efficiency in agriculture in Europe, and out of the debate regarding that form of production as a mechanism for surplus appropriation. However, since the turn of the century, with the evolution of the mode of production debate and growing discussions on the resilience of the peasantry, attempts have been made to apply and discuss the usefulness of the classical and neoclassical theories as well as Marxist views to an understanding of the role of the peasantry in Third World countries.

Classical and Neoclassical Theories

Adam Smith's discussions on sharecropping specifically referred to the practice of *metayage* in France; he noted that this tenure type no longer existed in England [Byres 1983: 35]. Following physiocrats, like Aristotle, he attacked the form of tenure as being inefficient and a disincentive to production, although metayage cultivation succeeded slave cultivation and was, therefore, a superior form of production. Smith considers it to be "inferior to fixed money rent tenancy" insofar as it destroyed the incentive to make productive investment on the land [Smith 1976: 389-91]. He traced the evolution of tenure types from slave cultivators to *metayers*, and finally to farmers, and perceived *metayage* as a "transitional" form of leasing which would be replaced by fixed money rent tenancy. However, his analysis concentrated on the development of land tenure arrangements from a purely economic standpoint [Smith 1976].

Smith attempted to influence the existing property laws, which he perceived as hostile to capital penetration, by tracing the evolution of defective leasing arrangements instead of tracing the alteration of property laws itself, and concluded that the alteration of property laws set in train mechanisms that modify contractual arrangements. It might be argued, on the contrary, that the existence of defective leasing arrangements did not necessarily imply the presence of defective property laws. The appropriate approach in analysing land tenure development would have been to "trace the alterations in property laws: and not, as Smith did, to interpret (or advocate) the change in laws by tracing what might appear to be defective leasing arrangements" [Cheung 1969: 33].

In the main, Smith focused on two major issues. First, "the impact of product sharing on labour effort and its implications for capital accumulation, and, secondly, on-farm investment" [Pearce 1983: 389-91]. He agreed that the emergence of *metayage* was consequent on the "desire to increase labour productivity", but its weakness, he noted, lay in its lack of stimulus for investment. Whereas *metayage* created incentives for greater productivity and more careful management of the labourer's time relative to slave cultivation, Smith contended, it did not motivate the cultivators to invest or maximize.

The West Indian experience, however, does not support Smith's thesis. The empirical evidence shows that the *metayers* were eager to accumulate. Their desire for independence from plantation slavery was the motivating force for production [Marshall 1963]. In addition, the *metayage* arrangement was different from the classical form that Smith analysed. In the St Lucian context, on-farm investment was undertaken by both planter and *metayer*, and the nature of investment varied in each phase.

John Stuart Mill [1848], contrary to Adam Smith, defended sharecropping (primarily the *mezadria* in Italy operating on half share) in Southern Europe. Unfortunately, he ignored the French situation in his defense and appeared to trivialize the central reason for sharecropping, namely the lack of capital for investment. From his information on the *mezadria*, he concluded that sharecropping was efficient given the prevalence of a general willingness on the part of the landlord to "provide capital for improvements".

The major weakness of Mill's argument lay in the source of his data, much of his information was provided by a resident proprietor. According to Sismonde de Sismondi (the resident proprietor), under the *mezadria*, the tenant had an interest in the property, "as if it were his own". There was substantive "accumulation of capital, invention of many judicious rotations, and the prevalence of a spirit of industry"... The collection of a numerous population, upon a space very limited . . . show plainly enough that this mode of cultivation [was as] profitable to the land itself as to the peasant" [Sismonde de Sismondi 1966: 41-42].

Largely because of its economic efficiency, Mill analysed sharecropping as an alternative to wage labour on large estates and argued that the economic efficiency of the practice rendered it a preferred alternative to wage labour and other tenure forms, such as a system of fixed money rent tenancy.

The neoclassicals, of whom Alfred Marshall was the leading proponent, expanded on the issue of productive or economic efficiency raised by the classical theorists relative to efficiency and optimality. His first formal statement regarding the allocative implications of sharecropping has come to be known as the 'traditional view'. He contends that, given the assumption of a 'mutual equality' (50:50 share of the produce) between landlord and tenant, and the provision for equitable division of the gross product in the contract, the tenant "will not invest resources beyond the level where the marginal cost of output is equal to half the value of the marginal product" [Byres 1983: 47]. The landlord will only invest if the marginal product is equal "to a minimum of twice the marginal cost". Consequently, the equilibrium level of resource allocation will be lower than that "under a system of owner occupation or fixed rents". Thus, he concluded that the total per hectare rent and total per hectare product will be below optimum.

Marshall was aware of the limitations of his model. He pointed out that the landlord could derive an advantage from the contract if he so desired, in spite of the predetermined shares. According to him, "whenever payments of all kinds made by the cultivator left him a margin beyond the necessaries of life for him and his family, together with those comforts and luxuries which were established by customs, the landlord was likely to use his superior strength to raise the payments in some form or other" [Marshall 1961: 40].

Regardless of the limitations of Marshall's formal model, it has become the 'bible' for traditional thinkers on sharecropping and the entering wedge for neoclassical economists. In spite of the 'refinements' to the model, the basic premise has remained unchanged, that is to say, sharecropping is essentially 'inefficient' relative to a system of free farmers in England or Denmark [Marshall 1963: 640-45]. Marshall, like Smith and Mill before him, attempted to rank several tenure arrangements according to some idea of economic efficiency [Cheung 1969: 44].

In the main, Marshall argues that share tenancy is inefficient because of the "disincentive effect of crop sharing on the tenant's work effort" [Otsuka and Hyami 1988: 31]. His criteria for the determination of productivity, efficiency and profitability is not altogether compatible with *metayage* in St Lucia (1840-1957). The share of the *metayer* ranged from one-half to two-thirds and then to four-fifths, over the three phases, thus offering the successful tenant a psychological impetus to increase his productivity. Insofar as the *metayer* stood to benefit more from the arrangement with each succeeding phase, his marginal product would likely be equal to or higher than his marginal cost (since he had the incentive to maximize).

From the neoclassical argument, it appears that the major issue surrounds the relative inefficiency of share tenancy at specific stages of capitalist development. Hence an absolute efficiency is not the issue. This, however, takes us back to our opening remark at the beginning of the chapter. The time dimension and context of the discussions (preindustrial Europe) were not intended to address specific agrarian conditions in Third World countries.

Both classical and neoclassical theories postulate that share tenancy is inefficient on two counts. First, it reduces the landlord's profitability. Secondly, it decreases the tenant's share of the return, either as a result of his lack of motivation to maximize, or the landlord's 'eating him out' unjustly. For the theorists, the puzzle seems to reside in the inability of their analyses to explain such factors as (i) customary share arrangement, which is not adequately explained by optimizing criteria, and (ii) the persistence and coexistence of sharecropping with cash tenancy and capitalist farming, in spite of its apparent inefficiency.

Marxist Views on Sharecropping

Marxist views have sought to fill the analytical gap in both classical and neoclassical theorizing on sharecropping. They contend that economics ought not to deal simply with production and management, but must also concern itself with the relations between persons or communities involved in the production process [Selsam and Martel 1977: 106]. It is certainly difficult to separate the social relation of production from the production process and yet capture the dynamics of change in production and accumulation, and in the wider society.

A central proposition of most Marxist conceptualization of sharecropping is that, under appropriate conditions, it is a particular system of surplus appropriation. That is to say, it is a form of land usage in which surplus labour is transferred to the landlord in the form of surplus product [Pearce 1983: 53]. The appropriate conditions are when capitalism is just beginning to penetrate agriculture. If the productive activity is characterized by unequal access to the means of production, the social relation is one of class; it is the class relation that provides the possibility for the transference of surplus labour, and the mechanism that facilitates this transfer constitutes a relation of exploitation. However, the form which the relation of exploitation assumed can be fixed or variable, and can be paid in labour produce or cash [Marx 1980: 1017-24]. A major part of the means of labour includes the extent of capital utilized, such as tools, machines, fertilizers, herbicides, and seeds. Control of the access to these elements can be a significant source of surplus labour. The nature of the distribution of the resources, then, determines the degree of dependence. The more unevenly distributed the means of production are, the greater will be the level of dependence of one section of the community for the reproduction of its existence.

Sharecropping in labour-scarce conditions allows the landlord to gain access to labour. The tenant owns the factor of production (labour), and requires means of production (land, capital and technology) "without which that labour cannot be utilized to the extent desired". The landlord, who possesses the means of production but no cash to pay wages, requires "labour over and above the quantity he is willing and able to supply" if these means of production are to be utilized. Therefore, those who own the means of production must get access to the labour of others [Pearce 1983: 53], particularly under conditions in which the labourer has not yet been alienated from the means of production.

Under sharecropping, the owner can use the contract as an instrument to acquire the labour of others and to provide him with some extra economic force (we will examine this issue specifically in chapter 5).

> Given that any form of contract will be based upon an inequality in ownership or control of means of production and/or labour power, it is the extent of this inequality and the degree to which one party to the

> contract can exercise influence over the other through it that will determine the choice . . . Where the distribution of productive resources is very skewed, the opportunity arose for a greater exaction of surplus labour by those owning those resources [Pearce 1983: 55].

Byres carries the discussion further than other Marxists who perceived sharecropping purely as a precapitalist economic formation. He pointed out that sharecropping must be seen

> in relation to the mode of production within which it is inserted, the 'laws of motion' of that modes of production, and the 'moment' of its development. In specific precapitalist/presocialist mode of production, it will represent for the dominant class, the most appropriate means of surplus appropriation: the means that will maximize labour input, land productivity, and surplus [Byres 1983: 10].

In essence, he put forward a case for the existence of sharecropping within underdeveloped and developed capitalism. Byres, like some others who claim to use a Marxist viewpoint, does not subscribe to the view that all feudal forms of production will disappear with the development of capitalism. Lenin argues that "in all capitalist countries, even the most advanced, there still exists survival of medieval, semi-feudal exploitation of the neighbouring peasants by the big landowners" [Lenin 1971: 598].

We can safely conclude that classical, neoclassical, and Marxist theories assume that sharecropping in its many forms will ultimately be destroyed (after a prolonged period of time, of course). While the classical and neoclassical theories attribute its destruction to its inherent economic inefficiency, Marxists ascribe it, instead, to historical factors, namely the development of large-scale capitalism in agriculture. For most Marxists, however, the replacement of sharecropping, a feudal remain, is inevitable, since it is a transitory form of production. The argument, therefore, is that all feudal forms of economic activities are ultimately destroyed in the process of capitalist development. Marx contended that "there should be succession of modes of production" and that the development of large-scale capitalist agriculture would transform peasant production—a precapitalist economic formation [Marx 1980: 21].

Marxist writers offered a more dynamic analysis of the form of production. They attempted to synthesize two crucial variables in the production process: the one, the social relation of production; the other, the notion of economic efficiency. In fact, one derives from the discussions appropriate tools to appraise both variables. In his study of the development of capitalism in prerevolutionary Russia, Lenin identified two specific roads through which capitalism could develop in agriculture—one evolving from the slow transformation of big feudal estates into capitalist enterprises, the other resulting from a variety of small capitalist farms [Lenin 1980: 17]. Lenin, in his analysis of the development of capitalism in Russia, discussed the emergence of semi-proletarian peasants, small, middle and big peasants, many of whom arose out of sharecropping with the decomposition of feudalism in agriculture [Lenin 1980: 592-96].

In the analysis of *metayage* in St Lucia, we are required to assess agrarian development, tracing the emergence, persistence, and development of an agrarian capitalist class over a specific historical period. It is in this sense that Lenin's discussion offers more relevant analytical tools in appraising the social processes than the neoclassical emphasis on relative economic efficiency. The analysis is expected to grasp both the economic and social processes at work.

Metayage: An Alternative Theoretical Framework

Theoretical considerations play an important part in virtually all aspects of scientific inquiry. They guide choice of variables, shape measurement procedures, and help interpretation of results. Theoretical reasoning helps to disentangle causation from association. Theory identifies patterns and ranges, and cause and effect. As Adelman and Morris [1975: 6] claim, it imposes a causal structure on empirical phenomena. Inasmuch as those European theories were not founded on the experience of agrarian development in developing countries, they might well be limited in providing an adequate analysis of the specificities of these countries. In fact, we propose that the treatment of *metayage* in the specific historical and socioeconomic circumstances of St Lucia required an alternative theoretical framework.

The set of propositions that follow is informed by two major hypotheses, namely:

1. *Metayage* provided the mechanisms for the emergence of an agrarian petty capitalist class
2. The nature and direction of rural transformation in St Lucia in general, and in Dennery, in particular, was conditioned by
   a. The character of *metayage* as a form of production
   b. The role of the emergent petty capitalist class in the economy and society

The theoretical framework is outlined under three broad propositions, namely:

1. Conditions likely for the emergence of *metayage*
2. Conditions likely for the persistence of *metayage*
3. The dynamic of *metayage* as a form of production

Conditions Likely for the Emergence of *Metayage*

Metayage is a French form of sharecropping apparently developed in the middle of the twelfth century in France with the "penetration of money and exchange into rural life". The *metayer* worked on the lord's land without any resource except his labour power, and depended upon his lord for "advancement of seed, provision of tools and even sometimes food for their families during the first year" [Byres 1983] and gave him one half of his produce in lieu of rent.

Although there is evidence of forms of production and labour process akin to *metayage* in the West Indies during slavery, *metayage* in the sense described above was not extensively practised in the British West Indies until the immediate postemancipation period. *Metayage* was first adopted in St Lucia about 1840, and after 1847 it spread to many other British West Indian islands. While it assumed a measure of importance in Tobago and Cariacou [Dookhan 1977: 13], it was only in St Lucia that this tenure type became an integral part of export agriculture until the shift from sugar to bananas in 1957. Although Dominica shares a similar rural culture with St Lucia, both having been colonized by the French, the practice never assumed that scale in export agriculture as it did in St Lucia.

The emergence of *metayage* in the British West Indies was conditioned by both exogenous and endogenous factors. The effect of exogenous factors was to throw the plantation system into serious crises and thereby create a need for new forms of agricultural production and labour usage. The endogenous factors promoted conditions ideally suited to sharecropping. The internal weaknesses of the productive process, both during slavery and in the period of transition to free labour, created the conditions for the adoption of *metayage*. A major weakness was its technological backwardness. Thomas [1988: 31] argues that "despite being integrated into a rapidly evolving world capitalist system, the planters remained remarkably precapitalist in their approach". Instead of trying to increase productivity by improving the efficiency of either the individual or the combined factors engaged in production, the planters invested more in land, slaves and buildings. The utilization of backward technology led to lower productivity and inefficiency [Marshall 1963].

In terms of the exogenous variables, developments in the global capitalist system and, in particular, on the British commodity markets created the conditions for the adoption of *metayage*. Among the major developments were the introduction of free trade in 1846 and the cash crisis in 1847-1848. The adoption of *metayage* cultivation, therefore, was part of the planter's attempt at consolidating capital use and gaining access to cheap labour.

As was the practice in Europe, in St Lucia, the planter utilized the contract as an instrument for securing his twin objectives of retaining labour and for deriving economies in production costs [Pearce 1983; Byres 1983; Marshall 1963]. He attempted and succeeded, to some extent, to restrict the mobility of labour and the bargaining power of the *metayers*. *Metayage*, therefore, was a precapitalist form of production underpinned by quasi feudal structural arrangements inserted within an underdeveloped capitalist mode of production. Under the system the form of accumulation was capitalistic. While the *metayer* was integrated within the wider capitalist system, he employed capitalist forms of production and accumulation along with other customary forms of labour usage. The arrangement, therefore, was inherently contradictory, synthesizing formal and non-formal mechanisms of production.

Long before the adoption of *metayage* in export agriculture, communal ownership, family ownership, and the practice of labour sharing were product sharing was prevalent in rural St Lucia [Thomas 1988]. Cultivation by permission was also an integral part of rural life. These types of land acquisition, ownership and cultivation were conditioned by French law and customs, which, up to the present, are the dominant cultural practices in St Lucia. Hence, the planters opted for a system of agriculture with principles that were familiar to the agrarian labouring classes.

Conditions Likely for the Persistence of *Metayage*

The commercial and fiscal policies of Britain with respect to her colonies in the nineteenth and twentieth centuries in a sense orchestrated the demise of the plantation system and created conditions favourable to the rise, establishment, and permanence of *metayage* and a *metayer* class. Protection for beet sugar and the passage of the Encumbered Estates Act in 1854, among other factors, together created conditions for the stability of *metayage* and the decline of the plantations (although the Encumbered Estates Act was intended to revive the sugar industry).

Official support for peasant cultivation also facilitated the stability of *metayage*. Whereas the British colonial policy tended to protect the planters and the plantation system in the nineteenth century, by the early twentieth century there was a significant shift in policy. Lobdell [1988: 202] noted that the British government encouraged the growth of the peasantry in the early twentieth century.

The strong sentimental attachment to the land and prevalence of the French customary form of labour organization (the *coud de main*—help from neighbours and friends for which no payment is made), and the strong land/home relation in the rural settlements served as support mechanisms for *metayers* and thereby facilitated the perpetuation of *metayage*. In Tobago and Cariacou, the internal financial and sociocultural support mechanisms that provided stability for the system in St Lucia were different. In Tobago, *metayage* cultivation was encouraged and supported by the merchant class and collapsed in the late nineteenth century with the decline of that class which had previously financed it. The system was extensively practised on the tiny island of Cariacou (one of the Grenadines) but failed to have wide-ranging impact on the wider social structure of the Grenadian society. In both cases there was a noticeable absence of that strong cultural attachment to the land which is still present in St Lucia. These propositions will be tested in chapters 4 and 6, respectively.

The Dynamics of *Metayage* as a Form of Production

Any inquiry into the Caribbean peasantries is likely to elicit much discussion, given the heterogeneous character of agrarian classes. In fact, "we lack anything like a complete account of the present distribution of peasant peoples in the islands" [Mintz 1974: 146]. These factors, therefore, present distinct analytical constraints in evaluating peasantries whose origins "have been diverse: some came into being only relatively recently; others flourished—and withered—at an earlier time" [Mintz 1974: 146].

Metayers: An Agrarian Social Class

Historians and writers on West Indian peasantries [Lewis 1936; Louis 1981; Acosta and Casimir 1985] have tended to treat *metayers* as an element or a stratum of the broad socioeconomic category, peasantry. Lewis, the pathfinder in the discussion of a peasantry in the British West Indies, argues that modern West Indian history began "without a peasantry", and, therefore, any attempt at analysing such a socioeconomic class in the West Indies involves "tracing . . . the birth and development of an entirely new class which has profoundly affected the foundation of West Indian society" [Lewis 1936: 1].

Lewis makes it quite clear that a West Indian peasantry does not manifest the characteristics of European peasantries. He defined the ideal typical peasant on the basis of the quantum of land he occupies. According to him, "the ideal peasant is perhaps a man who owns land enough to occupy his full attention, but not too much for him to cultivate alone, or with the help of his family" [Lewis 1936: 2]. The peasantry, he went on to explain, "covers a variety of tenures", and may include:

- peasants who are leaseholders
- peasants who are freeholders
- peasants who sell their labour in slack periods on their own holding
- peasants who own large enough holdings to hire labour occasionally to help with particular operations
- peasants who are involved in both agriculture and commerce [Lewis 1936: 2-3]

Although Lewis presents the peasantry as a heterogeneous class, he discusses its contribution as if it were homogeneous. While his analysis might be useful in explaining the dynamics of the peasantry at the macro level, it is limited in grappling with the phenomenon at the micro level, where the specific is not adequately representative of the general. To that extent, the analysis is inadequate for an understanding of specific peasantries, such as in St Lucia, where the specific does not conform to the general. The fact that Lewis makes no mention of the contribution of *metayers* as an agricultural social class leaves one to speculate about their location within the scheme of things.

In his attempt at tracing the origins of what he referred to as reconstituted peasantries, Mintz [1974: 146-56] showed that, within the Caribbean peasantry, there are a number of peasant types which exhibit specific characteristics. This underscores the importance of guarding against attempts at subsuming the particular into the general.

Louis [1981] has evaluated the struggles and contributions of the peasantry in St Lucia, but despite his application of micro analytical research methods and techniques, he arrived at the same conclusion as Lewis. He was probably not aware of a similar study by Marshall in 1963 on the role of the *metayers* in export agricultural production in the nineteenth century.

Theoretical and empirical evidence do suggest that *metayers* constituted a distinct social class in agriculture. Here we utilize the notion of social class consistent with Lenin's definition. According to him, "classes are large groups differing from each other by the place they occupy in a historically determined system of social production, by their relation, in most cases, fixed and formulated in law to the means of production, by their role in the social organization of labour and consequently by the share of social wealth of which they dispose". That is to say, a social class is defined by four major characteristics. The social class must:

- occupy a specific place in a historically determined social production
- be an owner or a non-owner of the means of production
- assume a role in the social organization of labour
- be determined by its dimension and mode of acquiring its share of wealth

As data presented later in the analysis will confirm, *metayers* in St Lucia occupied a specific place in a historically determined system of social production, and owned the means of production (labour). As sharecroppers, they were semi-independent, tied to the estates even if they operated on their own timetables. While they exerted little influence on the money and product market, they enjoyed a measure of influence on the labour market. They obtained their share of wealth through the share of their produce in cash and not in kind. The *metayers* manifested some characteristics of two of the four classes of peasants discussed by Lenin [1964].

Lenin [1964] referred to four classes of peasants in Russia—the semi-proletarian, the small peasant, the middle peasant and the big peasant. The semi-proletarian "till tiny plots of land . . . and obtain their livelihood partly as wage-labourers . . . and partly by working their own or rented land, which provide their families only with part of their means of subsistence". The small-scale peasantry as owners or tenants, "hold small plots of land which enable them to satisfy the needs of their families and their farms, and do not hire outside labour".

The middle peasants were those farmers who

(i) either as owners or tenants "hold plots of land that are also small but, under capitalism, are sufficient not only to provide . . . a meager subsistence for the family and the bare minimum needed to maintain the farm, but also produce a certain surplus which may, in good years at least, be converted into capital;

(ii) quite frequently resort to the employment of hired labour" [Lenin 1964].

The big peasants "are capitalist entrepreneurs in agriculture, who as a rule employ several hired labourers and are connected to the 'peasantry' only in their low cultural level, habits of life, and the manual labour they themselves perform on their farms" [Lenin 1980: 592-96]. The *metayers*, therefore, exhibited some similar features with the small and middle peasants.

One aspect of Lenin's conception of the peasantry is particularly useful to our analysis; that is, his conception of the agrarian social classes in relation to their involvement and integration in the capitalist system. Our discussion focuses on peasant classes involved in export agriculture between 1840 and 1957.

The Development of an Agrarian 'Capitalist' Class in St Lucia

The successful *metayers* were those who were able to take advantage of the opportunities for accumulation under the system. It was this mechanism of accumulation that was the vehicle for change, and, hence, led to the differentiation of the *metayer* class and the emergence of an agrarian capitalist class. The accumulation process is best grasped if the development of *metayage* is analysed in phases. Three phases are observable (i) 1840-1897, (ii) 1897-1929 and (iii) 1929-1957. The accumulation process varied with each alteration of the *metayage* system. In the period 1840-1897, the 50:50 share was the norm but in the subsequent phases, generally, the proportionate share of the *metayer* increased from two-thirds to four-fifths respectively. In each phase, the *metayer's* capacity to accumulate capital increased; his degree of dependence on the plantation lessened and his involvement in production for export, his relationship to the labour and money markets and his integration in the global capitalist economy deepened.

Contrary to established views [Marshall 1963; Louis 1981; Acosta and Casimir 1985], *metayage* had its peculiar dynamics and liberating mechanisms for the *metayers*. The nature of the accumulation process under *metayage* led to the development of heterogeneous social classes whose involvement in the global capitalist market was determined largely by their scope of production.

The foregoing provides the theoretical rationale for our first hypothesis. Our second hypothesis is derived from the first. Encapsulated, it is that the contribution of that agrarian capitalist class was instrumental in determining the pace and direction of rural transformation in St Lucia in general, and Dennery in particular. The character of classes changes over time. "They form, develop, and are modified by transformation in the society in which they exist" [Stavenhagen 1975: 27].

The validity of these hypotheses is our overriding concern in chapters 4 through 6. Chapter 7 appraises the utility of all the above propositions on the micro level. In our analysis, we are not necessarily "looking for cut and dried solutions, since we must . . . study the various planes, contradictions, possibilities of analysis of human exploitation, and the possibility of liberation" [Currie and Ray 1985: 581].

CHAPTER FOUR

The Establishment of *Metayage*

The Emergence of *Metayage*

Metayage cultivation was established "throughout the island of St Lucia as early as 1840, three years before it was introduced in Tobago, where it was practised extensively until the depression of the 1880s" [Hay 1884: 38].

Developments within the agrarian capitalist mode of production in the British colonies led to the adoption of the *metayage* system in some of the Windward Islands. Internal technical weaknesses that developed under the slave-based economy militated against productivity and profitability. As Thomas [1988: 31], has stated

> despite being integrated into a rapidly evolving world capitalist system, the planters remained remarkably precapitalist in their approach. Instead of trying to increase productivity by improving the efficiency of either the individual or the combined factors engaged in production, they invariably responded by merely acquiring more and more land, slaves, equipment, machinery or buildings . . . Although sugar production was a scientifically based activity . . . the planters . . . did not meet the requirements of the early application of technological innovation and advances developed elsewhere.

The "speculative and consumption-oriented approach" of the absentee planters caused the capitalist mode of production to be technologically backward and weak [Marshall 1963; Thomas 1988]. Up to the mid nineteenth century, steam engines, considered the most significant piece of equipment in effective sugar manufacture, made up only a small fraction of the manufacturing technology utilized by planters.

TABLE 4.1: MOTIVE POWER OF THE SUGAR FACTORIES, 1850-1865

Year	Steam	%	Other Motive Power	%
1850	16	19.3	62	15.4
1855	20	24.1	65	16.1
1860	23	27.7	144	35.7
1865	24	28.9	132	32.8

Source: *St Lucia Blue Book of Statistics*, 1850-1865.

As illustrated in Table 4.1, the percentage of factories utilizing steam engines increased by 9.6 while the percentage of those using other motive power increased by 17.4 between 1850 and 1865. It was this lag in technological application, contended Marshall [1963], which accounted for the decline of the sugar industry. With reference to the lack of technological innovation on St Lucian sugar estates, Special Magistrate Stanley[1] pointed out that the island was "too far behind others in agricultural improvements". The planters failed to utilize improved techniques even in areas where the terrain facilitated its application. Drysdale[2] noted that the district of Dennery was "the most backward in improvements", although "there [was] no land better adapted for the use of the plough" on the island.

Although the internal technological backwardness of the sugar industry was a major contributing factor to the introduction of *metayage*, commercial developments in the global capitalist economy in the mid nineteenth century facilitated its extensive adoption in the staple economies. The introduction of the Sugar Equalization Act in 1846 instituted free trade, thus bringing to an end mercantilism, the old system of protectionist trade for the colonial produce. British West Indian sugar economies were almost destroyed with the introduction of free trade since those high cost producing industries could not compete successfully with the low cost producing countries [Marshall 1963; Thomas 1988]. Those colonies that were more technologically advanced, like Trinidad, British Guiana and Jamaica, suffered less severely in the initial period but, ultimately, all became victims. Increased competition from foreign sugars resulted in an excess supply of sugar on the European market, thus causing prices to drop sharply. Low cost foreign producing countries were able to dump their sugar on the market at very low prices.

Table 4.2 examines the change in the pattern of consumption on the home market. The consumption of sugar and molasses from British possessions decreased from 2.7 percent in 1846 to minus 5.61 percent in 1847, while the consumption of foreign sugar increased phenomenally from 30.8 percent to 96.0 percent in the same period.

TABLE 4.2: QUANTITIES OF SUGAR AND MOLASSES IMPORTED FOR HOME CONSUMPTION (ENGLAND) 1844-1847 (TONS)

Year	British Possess.	% Change	Foreign Countries	% Change	Total
1844	4,145,044	—	93	—	4,145,137
1845	4,849,060	+14.5	34,584	+99.7	4,883,641
1846	4,985,792	-2.7	50,004	+30.8	5,045,856
1847	4,723,232	-5.6	256,424	+96.0	5,979,653

Source: C.O. 258/2 St Lucia Palladium, Dec. 3, 1853, p.3.

A cash crisis followed the commercial crisis in 1847. Many merchants who financed and marketed the colonial staples went bankrupt. As the planters became less and less able to service their debts, many banks refused to make further loans. Although statistics on the extent of indebtedness and encumbrances are not available, there is evidence from the official reports of its existence. The reports persistently referred to the high encumbrances that fettered the industry. For example, planters in St Lucia were heavily indebted. By 1850 mortgages against 43 of the 84 estates amounted to £200,000, £11,300 of that debt had been accumulated before 1834, £124,000 between 1834 and 1846, and £65,000 after 1846.

British investors lost confidence in the profitability of West Indian property. It was extremely risky to invest in West Indian property at the time, because (i) old debts had accumulated, and could only have been cleared with the unlikely sale of the estates; (ii) the diminishing profits of the industry militated against the certainty of receipts and interests. By then it was quite clear to the planters that access to any large capital investment into the sugar industry was no longer an alternative. "Under the pecuniary difficulties of the planters generally, the economy of manual labour, the adoption of new machinery, and implements" to reduce "expenses" was not possible. "The general state of embarrassment and poverty of the planters of St Lucia" and the "reluctance of capitalists . . . to make advances prevented any material improvements in the culture, and manufacture of the colonial staples". Although "immense fields of canes were planted" and often an abundance was reaped, the "enormous expenses" which attended them "absorbed the profits", and left "the planters hardly able to cover their expenses".[3]

Following the collapse of the sugar industry in 1846, the national income was being reduced at a rate of £10,000 annually until 1885.[4] By 1848 a wage reduction had been effected throughout the several districts of the colony. The daily wage of agricultural labourers ranged from 9 ½d to 1s 3d. Even the wages of overseers were cut as production levels declined and net returns dropped persistently. Payments of wages became so irregular that in many instances labourers were paid in kind rather than in cash.[5] As a result, the cash crisis led to a labour shortage in the industry, as many labourers preferred to squat on crown lands rather than to work for little or no money on the estates [Marshall 1963; Louis 1981].

Metayage was the most economic alternative. Efforts to attract large capital investment into the industry had proven futile; immigration to the Windward Islands was not an option, since the Colonial Office perceived such an alternative as "putting water in a sieve". Unable to meet the increasingly high production costs on the estates, and lacking the capital necessary to meet existing wage demands in the labour market, the planters opted for a temporary investment strategy until an expected upswing in the industry [Breen 1844: 301]. The planter who was "destitute of the requisite capital to enable him successfully to compete with his wealthier neighbour", took advantage of access to "the services of a few industrious cultivators, on condition of dividing the profits of the crop". The system allowed him to "derive a return, however scanty, from his land and the estate's machinery", which would have "remained idle and unproductive" [Breen 1844: 301]. The intention, therefore, was not to enter into a partnership but to remain in production.

The establishment of *metayage* in St Lucia was facilitated by important sociocultural and psychosocial variables. The principles of land sharing, product sharing, and land attachment, which were characteristic of *metayage*, were not at all foreign to the land-hungry labourers who perceived the arrangement as a symbol of their independence from plantation slavery [Marshall 1963]. Immediately after emancipation, noted Thomas, "the freed slaves acquired their lands either through communal purchase or individual proprietorships" [1988: 33]. Not only did the St Lucian small peasants acquire land collectively, but they also worked their plots through labour sharing *(coud de main)*. Family land is the most prevalent form of tenure among rural farmers and "is the collective inheritance of possession, use and usufruct rights in land by a group of persons who are related by blood" [Brook 1976: 30]. Therefore, the practice of land sharing, labour sharing and product sharing was already an integral aspect of rural agriculture before the adoption of *metayage*.

The official view in the mid nineteenth century favoured the adoption of *metayage*. Following the commercial and financial crises, both the metropolitan state and local capitalists saw sharecropping as the most rational alternative for economic survival. While the planters were solely concerned with the economic benefits that could be derived from the system, the state also perceived it as an appropriate mechanism for social and political stability. The Governor of Tobago explained that he was "disposed to attach an importance to the prevalence of Metairie System as conducive to internal security". He said this was a fact which was "altogether overlooked or disregarded". He argued that the discipline and industry that the system required would most certainly neutralize the possibility of social unrest ("riots and disorder") in a colony where "the emancipated negroes far outnumbered the other classes".[6] The *Port of Spain Gazette*[7] noted that the "labour force in St Lucia [was] indolent and restless", but attributed the state of restlessness to the "hot climate". While the available records do not indicate any evidence of widespread social insecurity in the colony that would have precipitated the adoption of *metayage* in St Lucia, Marshall [1963] noted that there was evidence of heightening tension between labour and capital over wages in the early 1840s. In fact Drysdale, in one of his December 31, 1848 reports, made reference to what he called "unsettled labourers" in Dennery.[8]

Table 4.3: Sugar Exports under *Metayage* (St Lucia) 1840-1865

Year	Production (Tons)
1840	1,883.3
1845	3,562.5
1850	2,695.1
1855	3,297.6
1860	3,647.0
1865	4,973.5

Source: W. K. Marshall, "Social and economic problems in the Windward Islands 1838-1865". p.42.

Metayage developed rapidly, and in a short time contributed significantly to agricultural export production. Table 4.3 demonstrates a steady increase in *metayage* production over the fifteen-year period. Production levels increased by 164 percent over the period. Access to Indian immigrants in 1856 would have contributed significantly to the increased production levels between 1855 and 1865. In 1847, two estates operating on *metayage* returned profits of £84 5s 11d and £482 5s 9d respectively, while four other estates operating on wage labour incurred losses varying from £337 17s 2d to £845 15s. The profitability of the two estates was attributed to lower production costs and higher productivity per acre. The yield ranged between 19 ¾ cwt. and 29 cwt. per acre for estates operating on *metayage*, and 19 ½ cwt. to 25 ¼ cwt. per acre for estates operating on wage labour. Costs of production were 16s 6d per cwt. on *metayage* and 16s 10d to £1 2s per cwt. on other estates [Lieutenant Governor's Report 1847]. These statistics do not support the view that *metayage* was "a make shift system, incompatible with a good system of agriculture" [Davy 1854: 284]. In 1841, the Stipendiary Magistrate for the fourth judicial district (Vieux-Fort, Laborie), referred to the system as "an excellent practice . . . not only worthy of adoption . . . in other districts" of the colony, "but also in other islands".[9]

In 1842, two years after its introduction in export agriculture, the system had become "an important and distinguishing feature in the history of agricultural progress in St Lucia".[10] Of the 19,881 cwt. of sugar produced in the 1947 crop, in the fifth district (Dennery, Praslin, and Micoud) 8,765 cwt. were produced under *metayage*.[11] In 1848, an estimated one quarter of the sugar cane plantations operated on *metayage*, and, by the late nineteenth century, the practice dominated the production of coffee, sugar cane, and cocoa [De Guiran 1986: 82].

The Use of the *Metayage* Contract

The benefits accruing to the planter or the *metayer* depended upon the advantages that the contractual arrangements allowed each party. There is evidence that the *metayage* contract was utilized as an instrument for appropriation. The planters used the contract for "securing the twin objectives of retention of labour force and economies in production cost" [Marshall 1963: 227]. Whether the agreements were written or oral, the provisions were generally similar, but with minor variations from estate to estate (see Appendixes). In the main, the conditions of the contract were as described in the Stipendiary Magistrate's Report of September 11, 1843:[12]

1. The planter provided about three acres of estate land to the *metayer* for about six years (the length of tenure varied) on which to grow, maintain, and cultivate a particular crop for a predetermined share of the produce.
2. The *metayer* was provided with his portion of land on the periphery of the estate on which to grow his provisions, raise cattle, and erect his house.
3. In some cases the *metayer* was given advances by the planter, which were deducted from gross produce at harvest time before the shares were determined.
4. The planter was responsible for all major capital expenditure, such as the provision of stocks, carts for transportation, and machinery.
5. The *metayer* was responsible for his labour cost and was given the value of his share of the produce in cash consistent with the current market price.
6. The planter reserved the right to superintend the cultivation of the canes and undertook the manufacturing of the canes.
7. The planter was given the right to take legal action against *metayers* for premature termination of contract. For neglect of duties certain penalties were prescribed.

The nature of the contracts on the islands on which the system was practised was conditioned by the specific structural, economic, and legal peculiarities of the different colonies. According to the Stipendiary Magistrate Reports, there were cases where the written contract appeared to have been deliberately made ambiguous and vague, to protect the interest of the planters. Given the incompatibility between *metayage*, on the one hand, and the legal regularities that governed land tenure, ownership, inheritance, and adjudication, on the other, and at the same time, the lack of legal expertise on the part of the stipendiary magistrates, *metayers* were left with almost no legal avenue for redress. In 1849, there were only 62 written contracts out of 645 agreements entered into.[13]

Initially, contracts were mainly oral because many *metayers* took precaution against affixing their signatures to what they perceived as instruments designed to reintroduce them into slavery. Oral contracts were more favoured because they offered the *metayers* an opportunity to avoid the heavy legal fees charged for the written contracts and allowed them to breach the agreement, in some cases with impunity. Those who favoured that kind of arrangement suffered great loss under the Encumbered Estates Act, since they enjoyed absolutely no immunity under its provisions. Passed in 1854, it worked disastrously for the colonies and those *metayers* who cultivated on verbal contracts. Until the Act was repealed in 1886, it provided for merchant creditors to have first claim on the estates to recover their debts. Some *metayers* who operated on contracts obtained very low returns for their produce, while the majority of those who worked on verbal agreements received no compensation when the estates were sold [Marshall 1963].

Despite the significant contribution made by the metayage system to the national product, *metayers* enjoyed very little protection under the law. In 1849 Governor Colebrooke penned his concerns:

> A Voluminous correspondence took place in 1845 and 1847, on the subject of Metairie contracts in St Lucia, . . . but . . . did not result in the adoption of the measures for the protection of the labourers in their engagements [Dispatch No. 40 of 1849].[14]

Marshall [1963] attributed the frustration of the colonial administration to the economic and political power of the planters, who, fearing that the legislature would establish the system, brought pressure to bear on the local legislators, thus forcing it to delay any attempt at ameliorating the conditions under metayage. However, by 1850, the legislature did succeed in passing an "Ordinance" to protect lessees and persons cultivating under the "Metairie System". The legislation provided for all contracts to receive the same legality as a lease and to be given a nine-year limit. It also provided for all contracts to be written in duplicate and executed in the presence of a stipendiary magistrate of the said district who was responsible for its registration. Disputes were to be settled in the local "Courts of Requests", and legal mechanisms were to be set up for proper adjudication.[15]

The available evidence is insufficient to allow for a rational discussion of the applicability of the assumption of some Marxist theorists that "the contracts were the landlord's instruments of exploitation" [Pearce 1983]. The absence of quantitative data, on the one hand, and the proliferation of verbal contracts, on the other, pose serious constraints. More importantly, the planter was different from his European counterpart, who was solely involved in share tenancy. He was a "capitalist/landlord", that is to say, his mode of reward constituted both rent (his share of the tenant's produce) and profits derived from whatever surplus he appropriated from a labouring class that exhibited "peasant-like means of production along with proletarian-like relations of production" [Frucht 1968]. The official reports strongly suggest that the planters were capitalistic and did utilize the contract as a mechanism of exploitation, particularly before 1850. The reports also indicate that the interest of the colonial officials was not always congruent with that of the planters. It was not the prerogative of the local administration to alter existing social structure. The primary role of the colonial government was to maintain law and order.

Accumulation Under *Metayage*

The *Metayer's* One-Half Share

The contract, which was utilized by the planters as an instrument of exploitation, also allowed the *metayers* certain advantages. There were clauses within it that offered the ambitious and industrious *metayer* opportunities to accumulate. The *metayer* was provided with enough land on which to build his house, cultivate his provision grounds, and raise his animals. Therefore, he had access to several sources of income, namely:

- income from the sale of his surplus from the provision grounds
- income from selling his labour power on the estate in slack periods
- his major income—his half-share from *metayage*

Planters and *metayers* alike appeared to have favoured the unwritten contract in order to maximize whatever advantage was available to them. Those informal arrangements, in some cases, allowed a *metayer* to occupy the land, concentrate on improving his provision grounds, and cut his productivity. The *metayer*, thereby, would reduce the planter's profitability, since the planter's marginal revenue was determined by the marginal product of all the tenants on the estates. In this case, both Adam Smith's [1976] and Alfred Marshall's [1961] argument concerning the derivation of the landlord's optimality would be useful, particularly in situations of strained relationship between planter and tenant. In such cases, the planter's marginal revenue would fall as a result of the decline in the *metayer's* marginal product. This is a crucial point, since the planter's profitability was dependent on the maintenance of high productivity levels from all *metayers* who cultivated the estate. The desired return on his investment required a sustained high marginal product on each plot. In such cases, the *metayer* could cut his marginal product on his plot without incurring any loss. He could maintain his desired return given his income from his provision grounds and from the sale of his labour power on the estate.

Contrary to established views that *metayage* was merely a system of exploitation [Davy 1854; Dookhan 1977], there is strong evidence to suggest that there was within the system an inherent liberating mechanism which made for the success of the industrious *metayer*. By 1841, *metayers'* accumulation in the fourth judicial district was significant enough to merit particular mention in the official reports:[16]

> On one estate in this district there resides a labourer who, cultivating canes in halves, planted during 1840 canes which will this year yield from four to five hogsheads of sugar. Taking, however, the lowest estimate—four hogsheads at the present prices of produce, each

> hogshead cannot clear less, after the deduction of all expenses, than £30; so here is a labourer earning by his industry £60 per annum . . . In addition, these canes were cultivated on Saturdays, holidays and during his spare time only, whilst all the rest of his time was spent working and receiving daily wages as a general labourer on the property, with free tenancy of house and provision grounds. His young children also . . . acquired habits of industry and knowledge whilst assisting their parents by carrying plants and manure to his cane piece.

An annual income of £60 was comparatively large for a rural small farmer. Wages ranged from 8d to 1s 3d per eight-hour day for an able-bodied labourer or a first class labourer [Marshall 1963: 180]. Using a median wage rate of 11s 5d per eight-hour day, a labourer working regularly on a five day week would have earned £12 10s annually. Therefore, apart from the other advantages of the *metayer*, he obtained at least five times the yearly income of a first class labourer for that year. In another district *metayers* cultivating 184 acres of canes produced over 285 hogsheads of sugar for an approximate share of £24 1s 6d.[17]

The returns from the system were so substantial for some *metayers* that they undertook the marketing of their own produce. Drysdale reported of a case in which the *metayer* refused to sell his eight hogsheads of sugar to the estate, but assumed the responsibility of shipping them.[12] The norm was for the *metayer* to leave all marketing arrangements to the estate and to receive the value of share subsequent to the sale of the sugar in England.

There were other reports of sizable gains from *metayage*. Table 4.4 below evaluates the level of accumulation for three districts for the crop year 1844. From the three districts, 41 estates with a total of 592 acres under cultivation produced 1,480,463 tons of canes under *metayage* cultivation, which brought the *metayers* an estimated gross return of £5,423 for the crop year 1844. However, we can derive additional tables from the above data to provide further analyses.

It is clear from Table 4.4 that the fifth district had the highest productivity per estate, but a smaller productivity per acre than the third district (Soufriere), where the number of acres under cultivation was about the same. Soufriere also had the highest productivity per acre among the three districts. Given that the norm was that the estate would supply one carre (about 3½ acres) of land to the *metayer* [Marshall 1963: 228], there would have been 52.6, 65.1, and 52 *metayers* in the third, fourth, and fifth districts, respectively, engaged in cultivation for that year. Therefore, the productivity per head in the third, fourth and fifth districts would have been 9,789.81, 8,686.93, and 7,692.31 tons, respectively. *Metayers* in the fifth district were the least productive, consistent with Drysdale's description of that district as the most technologically backward.

Table 4.4: *Metayer* Cultivation of Cane, 1844

District	Estates	Acreage	Qty Produced (Tons)	Estimated Value of *Metayer* Share
3rd	17	184	514,944	£2,060
4th	17	228	565,519	2,111
5th	7	182	400,100	1,252
Total	41	594	1,480,563	5,423

Source: C.O. 258/16, J.O. Drysdale, SM Report, Dec. 31, 1848, p.391.

Table 4.5: Level of *Metayer* Productivity (Tons)

District	Estates	Acreage	Prod.	Prod./Estate	Prod./Acre
3rd	17	184	514,944	30,290.82	2,798.61
4th	17	228	565,519	33,265.82	2,480.35
5th	7	182	400,100	57,157.14	2,198.35

Source: Computed from Table 4.4.

Table 4.6: Comparative Share by District

District	Estates	Total Share	Share/Estate	Share/*Metayer*
3rd	17	£2,060	£121.18	£39.16
4th	17	2,111	124.18	32.43
5th	7	1,252	178.86	24.00

Source: Computed from Table 4.4 and data from Marshall [1963: 228].

According to Table 4.6, *metayers* in the third districts realized the largest share of the returns, while those in the fifth district earned the lowest. In spite of the low returns, the average *metayer* in the fifth district earned more than twice the annual earnings of an able-bodied labourer.

The data strongly suggest that, prior to the commercial crisis of 1846, the labouring class was already being differentiated. A new propertied class was already emerging under *metayage*, although land was being deliberately sold at prohibitive prices in an effort to prevent the labourers from acquiring substantive holdings. Prices ranged from a low of £5 to £14 per acre in Grenada to a high of £18 to £20 per acre in Tobago. In St Vincent, it could have cost as much as £150 per acre while in St Lucia, it ranged between £2 and £20 per acre in 1841.[18]

Production levels also increased in the post-crisis period, and a few successful *metayers* derived substantive returns from their investments (see Table 4.3). In 1847, Drysdale referred to a "Return . . . from the 5th District" which appeared to have indicated that "the quantity of produce on the Metairie System had increased to 8,761,800" tons with an "estimated value of Metayers' share of £3,507".[19] The productivity per estate was 125,257.14 tons, and the productivity per acre was 4,817.58 tons. The productivity per head was 16,861.54 tons while the individual *metayer* share was £67.4. It is clear that 1847 was a 'boom' year in *metayage* cultivation in Dennery. *Metayers* earned more than twelve times the annual earnings of an able-bodied labourer in the pre-crisis period and twenty-four times the income of a first class labourer in the crisis period. The average annual earning of an able-bodied labourers was £12 10s. The daily wage rate was reduced by half in the crisis period, thus lowering the yearly income to £6 5s [Marshall 1963: 180]. In another district, not named, *metayers* cultivating 184 acres of canes produced 647 hogsheads of sugar and received a total share of £3,507 4s.[20] Since we do not know the number of cultivators, we are unable to determine the per capita share.

While the evidence presented is indeed very useful, it is not sufficiently comprehensive to allow one to generalize with respect to the precise nature of the accumulation process under *metayage*. A more definitive conclusion would require data on *metayage* cultivation over the entire period under review (1840-1897), that is, measurable indices such as productivity levels per acre and per head, value of planter's and *metayer's* share of the produce, value of capital investments of planter and *metayer*, market price of sugar, and so on. Notwithstanding these weaknesses, the available data strongly suggest that some *metayers* accumulated substantial amounts under the system.

The 50:50 share arrangement, the primary mechanism of accumulation, *ceteris paribus*, determined the returns that each party derived from the undertaking. There were many cases in which the *metayers* accused the proprietors of dishonest distribution of the proceeds of sugar because they were generally dissatisfied with lower returns generated from sharecropping [Green 1981: 255]. In spite of the numerous cases of fraud, evidence from the stipendiary magistrate reports confirms that some planters and *metayers* did adhere to the principle of 'mutual equality'. As the data demonstrate, the provision for equitable division of the produce did not deter some of the cultivators from raising the productivity levels. Those cases support neither Adam Smith's [1976: 389-391] nor Alfred Marshall's [1961: 44] assumption that the tenant cuts his marginal product so as to lower the landlord's profits. Although the classical and neoclassical theories were meant to appraise the specific conditions under *metayage*, it is clear that the specificity of the form of production underscores the limitation of their analytical usefulness. Given his small total outlay and access to cheap labour, the *metayer* could increase his returns without lowering his marginal product. Wages constituted the greatest portion of his capital investment, as whatever advances he received were deducted from his share of the produce at the end of the crop season. In most cases he was free from land tax, as only lessees and proprietors were expected to pay land tax. Most labourers became 'tenants at will'. The size and composition of the *metayer's* labour force depended on the size of his plot, his desired production level and the size of his family. Small *metayers* utilized family labour, while medium and large *metayers* utilized family labour, wage labourers, and helping hands (those who worked for no payment in *coud de main*). Therefore, given the *metayer's* small outlay relative to that of the planter, his access to cheap labour, together with his alternative sources of incomes, *ceteris paribus*, he stood to benefit more from the arrangement.

The *Metayer's* Savings

The small farmer stores wealth in a variety of ways. He may utilize the official financial institutions to lodge his savings. He may lend to a neighbour at low interest or convert his liquid assets into more fixed assets, such as investments in animals, precious metals, and land. Or he may simply hoard his money for precautionary and transactional motives [Firth and Yamey 1969]. In an agrarian society in which communication between town and country is at a low level, the level of illiteracy is high, and informal economic activity is substantial, official financial statements tend to underestimate the level of liquidity of the economy. Table 4.7 below shows the accounts of the Colonial Bank.

TABLE 4.7: DEPOSITS AT THE COLONIAL BANK, 1886-1895

Year	Depositors	Amt Deposited	Avg. Saving/Depositor
1886	296	£6,538	£22.01.09
1887	438	8,567	19.11.02
1888	453	9,492	20.19.01
1889	640	8,279	12.18.09
1890	681	9,492	13.18.09
1891	909	9,961	10.19.02
1892	905	7,704	8.10.03
1893	1,128	9,812	8.13.11
1894	1,187	9,428	7.18.10
1895	1,275	9,267	7.05.04

Source: Dispatch No.18 of 1896, Report on the Blue Book for 1895, Gouldsbury to Bruce. March 6, 1896, pp.10, 11.

The volume of savings mobilized by institution strongly suggests that the Colonial Bank did not serve as a banker for the planter and the merchant classes. Given the size of the annual average savings per depositor, it appears that a significant number of the working class within close proximity to the urban institution were willing and able to take advantage of its available banking facilities. The figures illustrate an interesting trend. There is a sharp increase in the number of depositors (over four times the number in 1886), a moderate increase of aggregate deposits over the period (41.7 percent), and a sharp fall in the average per capita saving (more than three times less). This trend suggests an increase in the number of small depositors. However, it is obvious that most labourers earning approximately £6 per annum (the average yearly wage during the crisis) would not have been able to afford the luxury of using the banking facilities. The historical evidence confirmed that the majority of the population employed in the civil service as well as many persons in the lower income group used the Colonial Bank as their depository.

The East Indian immigrants and rural small farmers constituted a significant part of the institution's clientele. The principal branch (the Castries branch) was opened in 1871, while two other branches were established in 1883, one each in Soufrière and Vieux-Fort. "The branch banks . . . were found to be total failures as depositors would not make use of them." It was discovered that "persons residing in the country districts preferred to bring their savings to Castries for deposit in the principal branch". The "East Indians formed no inconsiderable portion of the depositors", and in 1888 deposited £1,796 of the aggregate £8,567 deposited.[21] Therefore, the evidence does suggest that the clientele of the bank constituted a large number of small depositors, including many East Indian immigrants (who saved primarily for precautionary reasons) and *metayers* who were the largest accumulators (that is, the successful ones) among the rural population.

The historical evidence confirmed that labourers in St Vincent, Tobago, and Grenada were 'hoarding' their monies or investing in Friendly Societies for precautionary and speculative reasons. The pattern was the same for St Lucia, but the available data only provide information for the succeeding period.

In about ten years, the total value of the annual liquid assets of the bank rose from £6,538 to £9,267, that is, by 29.45 percent. The average annual saving per depositor fell from £22 1s 9d to £7 5s 4d (a drop of 67.10 percent). This may have been attributable to the depression that began in the sugar industry around 1880.[22] However, the official records noted that the level of transactions at the saving banks in St Lucia did not "reflect the general depression in agriculture".[23] From the foregoing discussion, the statement seems to suggest that those depositors in the agricultural belts were least affected during the depression. Therefore, we can conclude that the decline in the average annual savings per depositor was due to a shift in the transactional and speculative demand for money. As the following section shows, many *metayers* who were large creditors appeared to have done well, as they purchased whole or parts of abandoned estates. Some may have withdrawn large sums for that purpose. In addition, there were many large withdrawals by East Indian immigrants who repatriated large sums. In 1889, £1200 was withdrawn by "coolies [East Indians] who returned to India".[24]

Freehold Development

About 13,280 slaves were emancipated in 1834 [Augier et al. 1960: 183]. An estimated 2,797 of that population were persons of colour, who were mainly professionals, merchants, shopkeepers, and tradesmen.[25] Many of those who owned land were not given to agricultural production [Louis 1981]. In addition, most of the ex-slaves who were involved in subsistence farming obtained their land through communal purchase [Thomas 1988].

Table 4.8: Freehold Development in St Lucia, 1845-1897

Year	Freeholders	Labourers
Dec 1845	1,345	3,236
Jun 1846	1,388	3,541
Dec 1846	1,390	3,659
Jun 1847	1,329	3,926
Dec 1847	1,381	3,872
Jun 1848	1,848	3,845
Jun 1849	1,900	3,864
Jun 1850	2,180	4,231
Jun 1851	2,531	3,710
Dec 1851	2,280	3,807
Jun 1852	2,375	3,797
Dec 1852	2,398	3,942
Dec 1853	2,343	3,718
1897	6,000	na

na, not available.
Sources: M. Louis, "An equal right to the soil: the emergence of a peasantry in St Lucia 1838-1900"; *Royal Commission Report 1897*, p.81.

Small farmers and labourers who possessed the necessary liquid capital were most inclined to convert their monies into fixed assets in the form of land. Marshall [1963: 256] has pointed out that the chief ambition of the ex-slaves was "to possess freehold land which they regarded as the high road to full independence . . . But since a lack of capital often prevented its possession, *metayage*, as a semi-independent cultivation, became a stepping-stone towards the attainment of their objective." In 1841, "about 25 labourers in the fifth district purchased land for a longer or shorter period, varying from three to ten acres upon which provisions [were] . . . planted and huts [were] erected".[26]

Table 4.8 traces the growth of the peasantry in St Lucia. Between 1845 and 1897, the number of freeholders increased by approximately 4,655 peasants, an increase of 346.09 percent. According to the Royal Commissioners, the 6,000 peasants in the colony in 1897 "were formerly labourers on sugar estates who, under the *metayer system* purchased . . . plots of land with their proceeds of their share of sugar" (my emphasis).[27]

The 14.89 percent increase in the number of labourers over the same period is attributed to population growth resulting from natural increase and immigration. Some 2,844 liberated Africans, East Indians, Portuguese, and Madeiran immigrants were brought into St Lucia in the 1850s [Marshall 1963]. Of that number, 112 liberated Africans were distributed on six sugar estates in Dennery.[28]

In 1864, there were 28 farmers who owned plots of over 17 acres in size and cultivated a total of 1,952 acres of land in Dennery.[29] Many small proprietors took advantage of a land liberalization programme in the 1880s which was initiated by the colonial administration. The official records noted that 144 acres of land were sold at £1 per acre in 1887 while "there were 177 applications for Crown lands received in 1883" requesting "3,280 acres, representing 6,976 acres".[30] Some of those applications "dated back as far as 1883".[31] Plots of 64 acres in size were offered for sale in the Mabouya valley periphery in 1887. In 1888, with the deepening of the crisis in the agricultural industry, a number of *metayers* purchased whole or portions of estates, as many were put up for sale very cheaply. Large estates were being sold so cheaply that the national newspaper remarked that "Sugar estates are going for a mere song".[32]

Summary

With improved market prices for sugar in the 1850s, *metayage* became less economical to the planters, some *metayers,* and labourers. The planters and the *metayers,* "influenced by the same motive, were now led to a separation of interests". The planter found that it was "more easy and less expensive" to utilize wage labour, and the *metayer* discovered that it was "less difficult and more profitable to purchase land and cultivate it for himself".[33]

Whether we share the conviction that a number of peasant types existed prior to emancipation [Mintz 1974: 146-56] or subscribe to the view that modern West Indian history began without a peasantry [Lewis 1936: 1], it is indisputable that British West Indian peasantries are different in kind from European peasantries. Lewis' definition of that heterogeneous 'third class' strongly suggests that *metayers* were leaseholders. However, in his analysis of the lower stratum in the immediate postemancipation era, Frucht contended that the ex-slaves did not constitute a "pure social type". The labourer was neither a peasant nor a proletarian, but exhibited "peasant-like means of production" along with "proletarian-like relations of production" [Frucht 1968: 295].

The adoption of *metayage,* a precapitalist form of production, significantly altered the role of the social classes involved in agricultural production. The planter's form of accumulation was changed, but remained different from that of the feudal landlords. While one can argue that his share of the produce constituted rent, there is strong evidence to suggest that he utilized the contract as an instrument to maximize profits. The lower stratum became further stratified into two groups, the labouring class and the *metayers*. The labouring class remained wholly dependent on the estates for its main source of income, which it supplemented with subsistence farming on Crown lands, family land, or communally owned land [Thomas 1988]. *Metayers* constituted a distinct social class, which was partly dependent on the estate. The degree of dependence was determined by the quantum of land a *metayer* occupied, the amount of advances he received from the estate, and the fruitfulness of his provisions. The small *metayer* was more dependent on the estate, since he was both tenant and labourer; he still sold his labour power on the estate in slack periods. The large *metayer* operated wholly as a tenant and utilized family and wage labour to cultivate his plot. The small *metayer* cultivated the normal three acre plot while the large *metayer* cultivated as many as ten or more acres.[34]

The successful *metayer* received his one-half share in cash and not in kind (payment in kind was not the norm), competed for labour on the labour market, utilized non-formal labour arrangements (*coud de main*) as well as wage labour to maximize his returns from his investments and reinvested his profits—to purchase lands. By the end of the first phase of *metayage* (1840-1897), the *metayer* class exhibited marked differentiation. Many tenant/labourers, unfortunately, had fallen back into the labouring class, while the social position of some had not improved [Marshall 1963]. However, among the group of large *metayers* emerged "a class of small proprietors among the population" numbering "about 6,000", with "significant economic and political strength".[35]

There is evidence to suggest that the *metayers* developed a strong social solidarity, which they utilized to further their economic and political interest. "They were united, one willingly helped the other; they worked for each other under a system called *coud de main*, or help for which no money is paid." The formation of groups of 'helping hands' provided for them a secured labour supply at harvest time. The official records noted that *metayers* in Dennery "established a *Metayer* Association". One of its success stories was the cultivation of six acres of land for the purpose of meeting the "cost of a worship house and the expense of an *Agricultural School* for the children of the quarter (emphasis added)".[36] The small proprietors were debarred from participation in public administration. The franchise was characterized by high property and income qualifications that placed it beyond the majority of the citizens. Any extension of the franchise to the small proprietors was perceived as admitting "less educated and less intelligent voters" into the process of administration.

Metayage, however, had led to the "growth and accumulation of wealth amongst the labouring population" and "thereby increased the number of small farmers" [Breen 1844: 301-3] in the colony. The character of the accumulation process was the mechanism for social change.

CHAPTER FIVE

The Persistence of *Metayage*

Although *metayage* cultivation persisted in the sugar industry between 1897 and 1929, it never assumed its former prominence. Both endogenous and exogenous factors facilitated the stability of the system of share tenancy. The plantations gradually shifted towards manufacturing and became increasingly dependent on peasant production in the pre–Depression period. However, it was those developments in the global capitalist economy that constrained the declining capitalist planters to resort to sharecropping as a viable alternative.

The Consolidation of *Metayage*

Successive periods of crisis resulted in its expansion and the minimization of paid labour, while periods of boom brought about its contraction and the utilization of wage labour. The industry experienced a period of decline in the 1880s caused by the displacement of colonial sugar on the British market. The absence of British tariffs and a disguised export bounty allowed low-cost European beet sugar producers a monopoly of the market.

The data in Table 5.1 are quite revealing. While the consumption of refined beet sugar on the home market increased sevenfold between 1885 and 1895, there was a persistent decline in the level of export of cane sugar from St Lucia between the years 1893 and 1903. Export level dropped by 1,843,200 tons, a fall of 95.6 percent. Between 1893 and 1895 the importation

Table 5.1: Imports of Beet Sugar into the UK, 1885-1895, and Exports of Cane Sugar from St Lucia to the UK, 1893-1905 (tons)

Year	Refined Beet	Muscovado Sugar
1885	152,000	na
1887	311,000	na
1893	575,000	1,928,400
1895	706,000	763,700
1899	na	101,200
1903	na	85,200
1905	na	792

na, not available
Sources: *Royal Commission Report 1897*, Vol. 7, p.172; *St Lucia Blue Book of Statistics*, 1885-1905

Table 5.2: Sugar Prices on the UK Market, 1890-1895

Export	1890	1891	1892	1893	1894	1895
Mus.sugar (100 lbs.)	12s 6d	13s 3d	11s 7d	15s	10s 6d	8s 9d
Usine sugar	16s	17s 3d	14s 9d	19s	14s	12s 6d

Source: Dispatch No. 18 of 1896; Intercolonial Dispatches 1895-1897; Report on the Blue Book for 1895; Gouldbury to Bruce. March 6, 1896, p.43.

of refined beet sugar increased by 131,000 tons (22.78%), while the export of muscovado sugar from the colony decreased by 1,164,700 tons (60.4%).

Except with a few fluctuations, Table 5.2 points to a steady decline in the price of colonial sugar on the United Kingdom market from 1891 until 1895. The price of muscovado sugar which was produced in greater quantity, dropped by 46 percent over the five-year period. In fact, prices of cane sugar fell from 13s 3d per cwt. in 1884 to less than 10s per cwt. in 1897.[37] When "all the estates were in full swing (including peasant holdings) there were about 12,000 acres under cultivation; but in 1897 there were only 1,000 acres of estate land producing muscovado sugar, and about 2,500 under cocoa cultivation". Of the 86 estates on the island only 20 were in operation, while 66 were wholly or partly abandoned.[38]

Table 5.3: Status of the Sugar Estates in Dennery, 1897

Source of Energy	Estates	Avg. Crop (Hhds)	Actual Condition
Steam	Anse Canot	216	Abandoned
Steam	Fond d'Or	150	Contributor to DFC*
Steam	La Resource	60	Abandoned
Steam	La Caye	130	Contributor to DFC
Water	Mondesir	60	Abandoned
Water	Pearl	250	Abandoned
Water	Richfond	100	Contributor to DFC

* Dennery Factory Company Ltd.
Source: *Royal Commission Report 1897*, Vol. 8, pp.70, 71.

The situation in Dennery was not different. Although production continued in the district, levels fell sharply. Table 5.3 reports on the level of productivity on the estates. Only three estates remained in operation; they produced an average crop of 380 hogsheads of sugar annually. If the weight of a hogshead was the same as in 1857 when it ranged from 14 cwt. to 16 cwt. [Marshall 1963: 530], the aggregate production for the crop year would have been only 285 tons. It is also worth noting that Anse Canot, the only estate in operation in the highland area, was abandoned.

The depression in the sugar industry had adverse repercussions on the national economy. The year 1896

> was not . . . a prosperous one for the colony . . . Short crops, and low prices for produce, had their inevitable effect of narrowing the spending power of the people . . . revenue . . . fell short of the estimated revenue by £6,604 and the actual expenditure by £9,015, leaving on the transactions of the year a deficit or debit balance of £9,015 . . . In addition to short crops and low prices and indeed, arising out of those circumstances, there was a considerable reduction in the rate of wages paid to the labouring class and, of course, a correlative reduction of their spending capacity.[39]

The depression was accompanied by a persistent contraction in sugar production and rising prices of basic commodities and producer goods. That is to say, there was a reduction in the money supply and a fall in both nominal and real wages.

Regarding the cause of the depression, there were differing views. The president of the Imperial Agricultural Department traced it to the persistent utilization of backward technology. It was stated by way of elaboration that "the loss in actual manufacturing of the sugar amounts to one-fourth of the total quantity produced". By contrast, the planters attributed the economic depression to the bounty system, which they claimed prevented them from competing successfully with beet sugar producers, given the unequal terms of trade.[40]

The crisis of the 1890s was followed by a series of upswings and downswings in the global economy, each having its specific effect on the local economy. "The West Indies enjoyed a modest measure of prosperity" between 1900 and 1914 in "contrast to the extreme depression of the nineties" [*West Indian Bulletin* 1899: 490].

Although World War I caused dislocations in trade and commerce, the immediate postwar period brought about a brief time of prosperity. Excess demand for sugar caused prices to rise so appreciably that the average price of sugar per ton rose from a range of £10 to £20 to a high of £37 in 1920. However, by the end of 1929, the reintroduction of the bounty gave beet sugar producers a monopoly on the European product market, thus causing the price of cane sugar to drop sharply [Guillebaud 1930: 490].

The cycle of crises in the sugar industry facilitated the persistence of *metayage* cultivation, but on a less extensive scale. In that second phase, (1897-1929), the ratio of shares was no longer 50:50. A number of variations existed in the contractual arrangement between planter and *metayer*; but official reports noted that "besides giving one-third of his sugar the *metayer* got nothing of the molasses".[41] The *metayer's* share of the produce, therefore, had increased from one-half to two-thirds. In that phase the tenants also enjoyed much more bargaining power than those of the previous phase (1840-1897). According to one informant interviewed in May 1989:[42]

> The locataire and the estate manager would agree on the share . . . If the *metayer* was a good business man, and if his land was far from the rail lines, he would give one-third or even less. If the land was poor, . . . hilly or had no ratoons the locataire share was more according to their bargain.

The ratio of the *metayer's* share was determined by

- his ability to bargain with the planter
- the proximity of the land to access road or railway lines
- the price level in the commodity market at the time
- the fertility of the land at the time of acquisition
- the development of the land at the start of the contract

In some cases, where the *metayer* occupied marginal lands and did not have easy access to railway lines, he obtained as much as three-fourths of the produce, but those cases were few in the Mabouya valley.

The Decline of the Plantocracy

As sugar prices fell, cocoa assumed a position of importance alongside the other staples. Since cocoa fetched a higher price than sugar on the European market, the planters invested more of their limited capital in the industry. Table 5.4 documents sugar and cocoa production levels between 1899 and 1910.

Table 5.4: Production of Sugar and Cocoa, 1899-1910

Year	Sugar (cwt.)	Cocoa (lb)
1899	8,834,424	62,351
1900	8,612,320	1,298,000
1902	8,258,489	1,493,352
1905	3,721	1,893,600
1908	4,481	1,355,000
1910	5,199	1,637,400

Source: St Lucia Blue Book of Statistics, 1899-1910.

There is evidence of a simultaneous, persistent increase in the production of cocoa and an uninterrupted decline in the output of sugar in the period. Except for slight fluctuations, sugar production fell drastically from 1902 until 1910; production fell by 8,829,225 cwt., a 99.94 percent decline. On the other hand, cocoa production levels grew by 1,575,049 cwt., more than 26 times the 1899 level.

Cocoa was the major foreign exchange generator in the 1880s. In 1889, the 1,025,360 lb of muscovado sugar produced realized only £3,950, whereas £17,158 was obtained from the 953,700 lb of cocoa sold.[43] In fact, "cocoa prices reached a high of 128 shillings per cwt. in 1920", 88s over the anticipated 40s per cwt.[44] Cocoa as a tree crop facilitated the exploitation of labour under the *metayage* arrangement. "It worked to the great advantage of the estates, with the tenants benefiting comparatively little thereby . . . After the development of the lands, usually 4 to 5 years later, the estates retake possession allotting to the tenants other undeveloped (marginal) lands".[45] As soon as cocoa trees matured into full trees, the *metayer* was evicted and compensated in accordance with a fixed rate. In some cases whole estates were converted into cocoa plantations in that way.

Although some planters maximized their profits in the cocoa industry, their total revenue lagged behind their total cost, thus resulting in a loss on total investment. The planter's total fixed costs were so high that the marginal profits from his investment in cocoa could only delay his inevitable bankruptcy. In 1893 "estates comprising an area of 12,233 acres" belonging to the St Lucia Central Factory were "in the process of being escheated" because the management was financially incapable of repaying a £40,000 loan borrowed from the State's treasury.[46] By 1889 the central sugar factories (Dennery, Grand Cul-de-Sac, Roseau, and Vieux-Fort) owed the treasury a further "£33,100" payable at "1 percent per annum". The Sinking Fund "provided for the Central Sugar Factories" had been reduced "by £6,900".[47]

The planters no longer had the monopoly on capital and, as a result, they received stiff competition from small proprietors, who produced large quantities of sugar canes at lower costs (small proprietors utilized the *coud de main* to gain access to cheap labour). Sugar was manufactured in four central factories at Grand Cul-de-Sac, Roseau, Vieux Fort, and Dennery. "The output of those four aggregated 3,962 tons in 1897, valued at £55,497." The planters experienced difficulty in getting "a sufficient supply of labour," as many labourers were attracted to "military works", "to money to be made on the Coaling wharves", and to emigration to the "gold fields of Cayenne". In addition to the four central factories several small mills, driven chiefly by wind or horse power, were "maintained for the purpose of making muscovado sugar". In 1897 "exports amounted to 457 tons, valued at £3,900". The bulk of sugar was "grown by the factories, but much was contributed by peasant proprietors . . . Syrup, an article of favourite local consumption," was made largely in those mills.[48]

The antagonism between the planters and *metayers* (that is, those who acquired holdings) intensified with decreasing returns arising from competition in a contracting labour market. In addition, "as soon as he (the *metayer*) purchased a portion of land he had to set up a cattle drive". Every impediment was placed in the "way of the *metayer*" so as to prevent the "successful manufacture of his canes".[49] While the planter's economic power was crumbling the very few successful *metayers* and the emergent peasantry were rapidly gaining an increasing share of the agricultural industry.

Development of an Agrarian 'Capitalist' Class

Further Differentiation Among *Metayers*

By 1929, *metayers*, as a social class, had become further differentiated. One interviewee noted that there were a few *metayers* who employed as many as 24 labourers at harvest time, in addition to the labour of family members and helping hands; cane output ranged from less than one dozen wagons to as many as 30 wagons. There were other large *metayers* whose yield exceeded 40 wagons. The wagon weighed between one and one and a half tons, on average, when filled (depending on the skill of the packers as well as the type of canes produced).

Although quantifiable data were not available to verify the evidence gathered from oral history, the evidence suggests that *metayers* gained very little from *metayage* relative to their accumulation in the previous phase. By the pre–Great Depression period, two groups of *metayers* were clearly identifiable in export agriculture—those who cultivated cocoa and those who were engaged in sugar cane production.

The accumulation processes, as a result, were distinct, given the differential in crops cultivated and the conditions under which each group operated on the estates. The sugar cane growers were unable to maximize returns, due to price fluctuations on the world market and, more importantly, because of the increasing investment of estates in the cocoa industry. However, the two-thirds share arrangement did allow some fortunate *metayers* to accumulate substantial returns (we will discuss this issue shortly). Most of the *metayers* who cultivated cocoa were exploited and pauperized. The conditions under which they worked allowed the planters an effective mechanism for the exploitation of labour, insofar as the *metayers* were easily evicted from the estates once the cocoa plants were matured.

Metayage was the mechanism for accumulation and for upward mobility for the labouring class. It was from differentiation of the *metayer* class that the contributor (the small, dependent farmer) and the proprietor (the medium and large, independent farmers) evolved. The evidence suggests the existence of a reverse process of differentiation among the *metayer* class. A large number of the contributors seemed to have lost their potential for accumulation and were being proletarianized; that is to say, the majority of them were being relegated to the labouring class from which they had emerged. Although both sugar and cocoa *metayers* were victims, the cocoa *metayers* were the most exploited. In most cases the *metayers* were compensated after one or two harvests.

The empirical evidence underscores a major limitation of the classical and neoclassical theories developed to explain economic efficiency in European agrarian conditions. The assumptions concerning the *metayer's* manipulation of the marginal product and the marginal cost do not hold under such conditions of exploitation. Inasmuch as the theories do not emphasize the social relation of production, they cannot capture the dynamics of the incidence of unequal exchange under sharecropping. On the other hand, the Marxist paradigm is useful insofar as it assumes that sharecropping creates conditions for surplus appropriation and exploitation. Althusser and Balibar [1970: 212-16] argued that in conditions of unequal access to the means of production, mechanisms (such as the contract) are utilized for appropriating surplus. Pearce [1983: 53] contends that one cannot separate surplus labour from surplus product; the former can be transferred in the form of surplus product. In the cocoa industry the planter did succeed in using the contract for the twin objectives of gaining access to cheap labour and for exploiting the labouring class. Where the contract was written, it was so skilfully drafted, that the planter could have the *metayer* evicted and compensated very easily. In addition, cultivators were prohibited by law from selling cocoa without the permission of the proprietor.

Table 5.5: Annual Balances at Saving Banks, 1901-1929

Year	Depositors		Amt Deposited	
	Cast. Bank	Den. Bank	Cast. Bank	Den. Bank
1901	—	—	£19,182	£ —
1905	1,018	—	13,483	—
1906	1,371	—	11,376	—
1909	1,430	—	10,706	—
1914-1915	1,443	22	11,416	227
1916-1917	1,134	20	8,838	252
1917-1918	1,111	19	7,886	527
1918-1919	1,109	19	7,368	874
1920	1,076	33	7,329	647
1921	1,013	33	3,696	184
1929	517	38	1,821	137

Source: *St Lucia Blue Book of Statistics*, 1901-1929.

Perhaps an indication of the level of accumulation of those *metayers* who cultivated sugar cane in that period is provided by the financial records at the Dennery branch of the Colonial Bank for the period under study. Table 5.5 portrays a general decline in savings in the form of liquid assets in the banking system, perhaps indicative of the conversion of liquid assets into fixed assets. The number of depositors in the Castries Colonial Bank fell by 49.2 percent in 28 years, but increased by 72.73 percent in 15 years at the Dennery branch. Wartime savings decreased by 35 percent at the Castries branch, but increased by 72 percent at the Dennery branch. The per capita deposit at the Castries branch in the period was £9.2, while the per capita deposit at the Dennery branch for the same period was £15.5. Small proprietors and *metayers* were the main patrons of the bank in Dennery. It is almost certain, however, that the majority of creditors were small proprietors, given that the number of depositors increased by only 16 in the 15-year period. In addition, evidence shows that some *metayers* accumulated fixed assets in the period preceding the establishment of the branch of the Colonial Bank at Dennery. In 1907 *metayers* acquired approximately 667 acres of cultivable lands in Dennery.[50] Although data on the growth of freeholding in the period are very scant, available colonial reports noted that in 1909 large areas formerly under cane were abandoned and purchased in whole or in part by successful *metayers* and proprietors.

Table 5.6: Accounts of Friendly Societies in St Lucia, 1899-1913

Year	Number	Membership	Receipt	Disbursement
1889	6	491	£405. 3. 4	£341.16.9
1900	7	638	35. 3. 7	541.11.0
1902	9	879	721.12.11	1,286.16.9
1905	12	1,247	935. 1.10	1,778.12.4
1908	16	1,526	1,707. 7. 9	1,668.18.2
1912-1913	24	2,690	3,588. 0. 4	2,461.18.2

Source: *St Lucia Gazette* 1899-1913

The financial records of the savings bank could not provide an adequate measurement of wealth in the rural areas, since the Friendly Society was another major, semiformal financial and welfare institution among the rural population. The Society offered its beneficiaries "funeral expenses" and sicknesses benefits.

Table 5.6 illustrates the persistent growth of the welfare institution in the colony. For the period under discussion, the number of institutions quadrupled; membership expanded fivefold to 2,600, and total receipts rose from £405.3.4 to £3,588.0.4 (about nine times the value of its initial receipts). The rapid growth of the Friendly Societies in the colony indicates the level of confidence in and identification with the institution, and reinforces the fact that the rural population had knowledge of the precautionary demand for money apart from the transactional and speculative motives that were discussed previously.

The Growth of an Emergent Propertied Class

By the turn of the century the emergent peasantry was gradually becoming more and more involved in the network of agricultural production in the colony. "A considerable amount of cane" was obtained from peasant farmers, numbering 744 for the four factories".[50] Despite the fact that capitalist banks refused to provide loans at moderate interest rates to the peasantry, the small proprietors were "the only set of planters to whom it" was "safe to make loans". Their "small plots of land which they could give as security" were "unencumbered".[51]

The degree of involvement of the emergent propertied class who became known as cane farmers (that is, those medium proprietors who specialized in cane production), depended upon the size of their holdings, the amount of working capital they had access to, their access to labour, their competitiveness on the product market, and their access to technology. It is possible to identify three distinct groups of cane farmers involved in export agriculture, if they are classified according to their degree of independence and involvement in the production process. There were the medium cane farmers, who cultivated several acres of land, employed large gangs of wage labourers, operated "small mills", and worked part of their lands on sharecropping.[52] They operated independently of the estates and sold their produce to the central factory at the current market price. One interviewee who worked for one of those medium cane farmers recalled that "there were some people with large pieces of land . . . twenty or thirty acres" who gave their "friends or neighbours" portions to cultivate on "share for one-third or one-fourth" of the produce.[53]

The small cane farmers became contributors under the *contributor system*. Under this arrangement, the planter furnished those small proprietors (who lacked the working capital) with advances, on condition that they sell their cane to the estate at a prearranged market price. The contributors could not alter their selling price, even if the market price of the canes appreciated at the end of the crop season. Of the 6,900 labourers employed in the sugar industry between 1929 and 1930, 900 were *metayers* and contributors.[54]

Small proprietors were differentiated in terms of access to capital and degree of dependence on the estates. By the Great Depression, the social classes in the export agricultural sector had become further differentiated. In, hierarchical order, they constituted:

- planters who owned the estates
- small and medium peasants who operated independently

- contributors who cultivated their own plots, but depended on the estates for working capital and sold their produce to the estates at predetermined prices
- *metayers* who cultivated sugar cane
- *metayers* who cultivated cocoa
- labouring classes who sold their labour power on the estates but also squatted on Crown lands

There is evidence to suggest that, with the falling price of sugar and the simultaneous rise in the price of cocoa, many small and medium farmers injected part of their capital into cocoa cultivation to minimize their losses. Out of a total of 1,315 peasants engaged in planting cocoa and sugar, 1,187 possessed holdings less than 10 acres, while 8 had holdings of over 50 acres in extent. In 1905-1906 the export of cocoa was valued at £41,520.[55] The official reports noted that approximately 667 acres of cultivable lands changed hands in the district of Dennery in 1907, but details concerning the transactions were not provided, except for the names of the purchasers. However, we can only assume that the medium cane farmers were the most likely buyers, insofar as they constituted the most financially secure class.[56]

Data on the nature of the accumulation process and, in particular, statistics on the quantum of wealth accumulated by each agrarian social class in this period are deficient. Apart from the absence of statistics on the trend of production quantums and returns from investment for the emergent peasantry, there is a lack of concrete evidence on their involvement in the partisan politics. But it is quite recognizable that the peasantry was no longer a marginal factor in the agricultural sector as well as in the global capitalist economy. Thomas [1984: 124] noted that, on their visit to Guyana in 1897, the Royal Commissioners "recommended that peasant cultivation of sugar cane should replace the plantations, with the plantations confining their activities to processing". While there is no evidence of such recommendation for St Lucia, there was a noticeably increasing dependence of the estates on peasant production. *Metayers* and contributors produced a significant share of the estate sugar cane. One official report strongly suggested increasing peasant control of the sugar industry:[56]

> This is largely due to the fact that since *the disappearance of European proprietors* who cultivated large estates, *the greater number of cultivators are peasant holders* (emphasis added).

It is evident from the record that the capitalist proprietors were gradually losing their monopoly; the planters were on the decline, while the peasantry was on the ascendancy.

Constrained by economic difficulties at home and their inability to inject the required financial support into the colonial sugar industry, British officials favoured the increasing role of the peasantry in the industry.

> Throughout the period 1850-1883, most British officials continued to accept the planters' view that West Indian social and economic progress was endangered by the growth of independent peasantries. But in contrast with early views, official opinions after 1897 generally argued that the economic crises experienced within the West Indies had not been caused by the peasantry. On the contrary, if such crises were to be overcome, the peasantry would have to be encouraged and strengthened [Lobdell 1988: 198, 202].

The economic strength of the peasantry provided for them a base for political power. For the first time they began to influence the political direction of the colony at the administrative level. Representative government was introduced in 1925, and candidates came forward as 'independents'. However, because of the restrictive franchise, "there were hardly 50 names on the voters' list in the main towns; an elector was required to have an annual income of not less than £50 or property of equal worth". The voting age was 21 and 30 for men and women, respectively. Among the many qualifications of the candidates, they were required to have freehold or leasehold valued at least £150 above all charges for encumbrances.[57] This restrictive qualification, indeed, debarred a significant number of peasants, but the few who were eligible won comfortably. "The choice between the candidates [went] naturally and normally to the candidate with more property and friends in the district."[58] This, perhaps, traces the genesis of clientelism in electoral politics in St Lucia. Given the land culture in the colony, the electorate might have supported the medium peasants as an expression of their "reciprocal loyalties based on expectations of mutual benefits" [Lemarchand and Legg 1972: 152]. Their 366 electors elected three candidates who presented themselves for the Northern, Eastern, and Western constituencies unopposed in the 1928 election.

Summary

For the first time in the history of agricultural development in the colony, a new social class from the labouring population exercised a substantive measure of control of the sector. By the early 1920s, the number of European proprietors who cultivated the large estates had decreased significantly, and a greater number of peasant holders were involved in export agriculture.

The peasantry, which was gaining increasing influence in the agricultural sector, had become more differentiated, and more integrated into the global capitalist economy. There were three distinct socioeconomic classes involved in export agriculture. First, there were the *metayers,* who were semi-independent and cultivated the planters' lands on shares. They were non-owners and assumed a marginal role in the social organization of labour. Their mode of acquiring their share of wealth kept them away from direct involvement in the product market. Secondly there were the *contributors,* who were semi-independent, but enjoyed a greater degree of freedom than the *metayers,* insofar as they owned land, but lacked working capital. They were also more influential in the labour market than the *metayers,* and acquired their wealth through involvement in the domestic market. However, they lacked competitive strength on the product market, being constrained to sell their produce to the estate, which provided them with the working capital, to which they were contracted at prearranged prices. Thirdly, there were the *medium peasants* who operated independently of the estates. They possessed the means of production (capital, land, and technology) and played a major role in the organization of labour. They were involved both in the domestic and foreign markets and accumulated their wealth through profits and shares.

The quasi feudal structural arrangement that underpinned the underdeveloped capitalist mode of production, along with customary forms of production and circulation utilized by the rural farmers, gave colonial agricultural production a character that was unlike the European agrarian production discussed by the neoclassical theorists and the Marxists.

The absence of sufficient quantifiable data on production and accumulation makes it extremely difficult to ascertain the existence and the extent of surplus appropriation under *metayage.* However, the available evidence does suggest that, in the cocoa industry, the planters succeeded in utilizing the contract as an instrument for surplus appropriation.

The peasantry in St Lucia had neither the means of production nor the relationship to the means of production that characterized Karl Marx's European capitalist class [Marx 1968]. Notwithstanding the difference between the two groups of peasantries, there are certain peculiarities in the development of the St Lucian peasantry which are similar to those of the Russian peasantry (which is analysed in the works of Lenin) during the period of the decomposition of feudalism in Russia [Lenin 1980: 593, 596]. In both cases, there is an emergent 'capitalist-peasant' class. While in the case of Russia, Lenin identified four peasant classes—the semi-proletarian, the small, the middle and the big capitalist-peasant—in the case of St Lucia two social classes can be identified: the small and the medium (or middle) peasants. While these groups have some similarities with Lenin's categorization, the evidence suggests that they manifested varying degrees of capitalist characteristics. A distinct difference between the Russian peasants and the local peasants lies in the fact that the medium peasant class manifested more "capitalist means of production" and "capitalist relations of production" than the small peasant class. Lenin pointed out in the differentiation of the peasantry in Russia that there emerged a "rural . . . well-to-do peasantry" from which emerged a "well-to-do peasant class of capitalist farmers" [Lenin 1964: 174]. A somewhat comparable process is observable in agrarian development in St Lucia.

CHAPTER SIX

The Final Phase of *Metayage*

In the first half of the twentieth century *metayage* was the only economic option for many planters in a period of fluctuating fortunes in the sugar industry. Competition between cane sugar and beet sugar re-emerged and, between the periods 1919-1920 and 1924-1925, the yearly world production of beet sugar increased from 3,326,000 metric tons to 8,316,000 metric tons, giving the European beet sugar producers a monopoly on the European market. In 1928, reduction in production costs and preferential treatment on the British and Canadian markets injected some vibrancy into the colonial sugar industry. Wartime (1939-1945) dislocations brought an initial period of recovery, producing a situation of under supply, which caused prices of cane sugar to double. However, by 1953 surplus supply brought prices down to their lowest since the war.

The Decline of *Metayage*

Although *metayage* did not enjoy a position of prominence in the post–Great Depression period, it persisted in St Lucia and Cariacou. In 1931 there was no wage labour on the tiny island of Cariacou; all cultivation was by the *metayage* system.[59] In St Lucia, although the *metayers* remained an important producer in export agriculture, the number in production was on the decline. Of 750 cane farmers in Dennery in 1929, only 150 were *metayers* cultivating sugar on 130 acres of the Dennery Factory Company lands.[60]

The Rise of Peasant Agriculture

The accumulation process in export agriculture between 1929 and 1957 was conditioned by the fluctuations in the sugar industry, the changing character of sharecropping, and the increasing involvement of the peasantry in export agriculture.

Although the contract offered the *metayer* a greater share of the produce than was offered in the previous phase (1897-1929), there was "seldom any written contract". In spite of the fact that the Civil Code of St Lucia (see Appendixes) provided no security for non-formal tenure types, the *metayer* preferred the verbal agreements. "The lease of a plantation or farm, when no term was specified, was presumed to be an annual lease, terminating on the first day of August each year".[61] The legal provisions were still in the planters' favour, but by that time he lacked the financial, economic, and political power required to take full advantage of the exploitative mechanisms available to him. The Colonial Report of 1951-1952 noted:[62]

> on the sharecropping system the owners of the land generally demand one-third of the proceeds, the tenant bearing all the expenses for clearing and cultivation. Sometimes, however, the owner supplies part of the planting materials. In very rare cases the tenant receives all the proceeds from the land on the agreement that he plants, on the land rented, such permanent crops as coconuts, cocoa, and coffee.

The Report of the Commission of Enquiry into the Stoppage of Work at the Sugar Factories in 1952 reported that *metayers* gave one-fifth of the canes supplied to the estates in lieu of rent[63] and not one-third as is stated in the Colonial Office Report. For the purposes of this analysis one-fifth is accepted as the norm, insofar as the commissioners (who were nationals) were more knowledgeable concerning the practice and, more important, their information was derived from primary sources (the enquiry was done on the estates, and both managers and *metayers* appeared before the commissioners).

The successful *metayer* derived a larger share of the produce than was obtained by the tenants in the previous phase (1929-1957). When the crop was ready for harvesting the factory claimed one-fifth of the canes, and purchased from the *metayer* the remaining four-fifths at the current market price. The conditions of the contract, then, allowed the *metayer* a greater degree of independence and involvement in the domestic product market, in comparison with the previous phase.

In addition, a few cocoa *metayers* cultivated rent free. In such cases, the planters must have opted to forego their benefits in the short run for a larger return in the long run. The rationale would have been economical insofar as the *metayer* was evicted without compensation once the tree crop became mature trees, or after one or two harvests. But the contradiction in the arrangement lay in the fact that the cultivator freely accumulated from the land between three and five years.

By the end of the 1930s, the cocoa *metayers* constituted a marginalized class in export agriculture. Tropical diseases, unfavourable climatic conditions, and price fluctuations forced the planters to shift their investment from the cocoa industry to more viable industries. In 1938, some 6,198 cwt. of cocoa were produced at a value of only £8,332, compared with 5,585 cwt. sold in 1937. By 1938 there were approximately 3,000 acres, 4,500 acres, and 850 acres of cultivable land under sugar, cocoa, and bananas, respectively. Banana was perceived as the best alternative export crop.[64] The cocoa *metayers* had been so marginalized that they were no longer being mentioned in the colonial reports of the post–Depression period.

Table 6.1: Number of *Metayers* and Contributors in Cane Production, 1953

Area	No. of *Metayers*	No. of Contributors	Total
Roseau	582	195	777
Cul-de-Sac	563	140	703
Dennery	312	298	610
	1,457	633	2,090

Source: *Progress Report on the Agricultural Development Programme for St Lucia for the Period Ending June 30, 1954*, p. 19.

Cocoa cultivation continued on a low scale on some peasant holdings and on some estates, but more attention was placed on banana as a foreign exchange earner. Although production levels of cocoa increased from 4,032 tons to 4,385 tons, the gross aggregate value fell from $321,903 to $190,219 between 1954 and 1956.[65]

Some of the sugar *metayers* benefited substantially from the system and, as a result, they became further differentiated as a class. Together with the contributors, they produced a significant percentage of plantation sugar production. This is suggested by Table 6.1.

The evidence suggests that a minimum of 3 acres of estate lands were allotted to each sharecropper. If we were to use this figure as the average farm size, we could conclude that there were no less than 4,371 acres of estate lands under sharecropping in 1953. Therefore, the 1,457 *metayers* cultivated 4,371 acres of estate lands. Moreover, there were 633 contributors, semi-independent farmers who supplied sugar cane to the estates. Hence, the evidence strongly suggests that the estates were heavily dependent on peasant (*metayers* and contributors) production.

As portrayed in Table 6.2, *metayers* and contributors (peasant farmers) supplied 18,879 tons (9,599 tons and 9,280 tons, respectively) of the 100,084 tons of canes produced by the plantations, that is, 18.86 percent of the total production for the year. The fact that peasant farmers at Dennery contributed 36.7 percent of the production of the estates for the year seems to suggest a greater involvement of those peasant cultivators in commercial agriculture.

Table 6.2: Sugar Cane Produced in 1951 by *Metayers* and Contributors

Depot	Total Manufactured	Contributors Amt	Contributors %	*Metayers* Amt	*Metayers* %
Roseau	44,867	1,279	2.8	4,267	9.5
Cul-de-Sac	32,479	2,336	7.2	2,645	8.1
Dennery	22,738	5,665	24.9	2,687	11.8

Source: *Report of the Commission to Enquire into the Stoppage of Work at the Sugar Factories in St Lucia, in March 1952 and into the Adequacy of the Existing Wage-Fixing Machinery in the Colony*, p.13.

The data strongly suggest that the labouring class in Dennery were most involved in export production and therefore were the major beneficiaries of the system. The fact that the number of contributors in Dennery was almost three times the number in both Cul-de-Sac and Roseau reinforces the argument.

There are two principal explanations for the greater integration of the peasantry in export agriculture in Dennery. First, *metayage* had been most extensively practised in that district. Secondly, the planter preferred to increase labour intensity than to raise productivity through technology application (see chapters 4 and 5). In the main, the evidence does suggest a growing involvement and influence of the peasantry in commercial agriculture. With the declining price of sugar on the world market, low yields, and high cost of production, the estates were unable to meet increasing wage demands in the industry.

Those *metayers* and contributors who were contracted to the estates obtained very low returns from their investments: "In 1951 the price paid by Dennery Factory Company for contributors and *metayers* . . . was $7.24

per ton." This amount was 48 cents per ton less than the price paid by the other two companies (Roseau Company Ltd and Cul-de-Sac Company Ltd). *Metayers* and contributors bought fertilizers at 4s 5d from the company but, on the average, produced no more than 10 to 15 tons of cane per acre. Therefore, they operated at a loss, selling their produce at $7.24 per ton.[66] In 1951, all *metayers* engaged in export production generated a gross income, in 1951, of $79,095.76, while those in Dennery earned $19,453.88.

For the financial year 1951, contributors and *metayers* operating in Dennery, the area in which production was concentrated, contributed $146,282.96 to the colony's foreign exchange generation and $60,468.48 to the rural economy. Contributors in Dennery generated more money ($41,014.60) than their counterparts at Roseau and Cul-de-Sac, who obtained $92,259.96 and $16,913.64, respectively. The gross earnings of the *metayers* in Roseau were the largest—$30,893.08, compared to $19,149.80 and $19,453.88 for those in Cul-de-Sac and Dennery, respectively.

It is clear from Table 6.3 that the *metayer's* individual return from his investment was much lower than that of the contributor. Insofar as the above figures represent gross earnings, the *metayer's* four-fifths share (that is, average individual share) would have been $42.46, $27.20, and $49.88 for *metayers* at Roseau, Cul-de-Sac, and Dennery, respectively. His net return would have been lower, since he would have had to deduct his total cost of production from his gross returns.

Table 6.3: *Metayer* and Contributor Earnings from Sale of Sugar in 1951 (in Dollars)

Estate	*Metayer*	Contributor
Roseau	$53.08	$207.88
Cul-de-Sac	34.01	153.03
Dennery	62.35	135.63

Source: Computed from Tables 7.1 and 7.2 using additional data from the sources utilized for the same Tables.

An able-bodied man earning $1.06 per eight-hour day in 1951[67] and working for eight months (the normal duration of the crop season in St Lucia) earned more ($169.60) than all the *metayers* and most of the contributors. However, the analysis is limited on two counts. First, the *metayer's* income from sugar was only part (although a large portion) of his total earnings. Some *metayers* obtained additional income from selling their labour power on the estates, and from sale of surplus produce from their provision grounds. Secondly, the utilization of the average net income of all the *metayers* does not provide an adequate analysis of the actual distribution of earnings. One *metayer* from Cul-de-Sac, who appeared before the Commission in 1952, testified that he lived in his own house on the company's land and produced 83 tons of canes in 1951. He utilized family labour and "helping hands" for cultivation and, in addition, employed paid cutters and headers at harvest time. He earned a gross of $640.76 in 1951, and stated that he was satisfied with his returns. However, when questioned about the remuneration of his labourers, he replied that the labourers knew "he was poor", and therefore did not "expect to receive the same rate [offered] by the company".[68] The *metayer*, it seems, was either bent on getting the most out of his labourers for less, or benefiting from the practice of labour sharing, which was part of the rural culture. Another *metayer* from Dennery, Joseph Sardusingh, who appeared before the 1952 Sugar Enquiry, admitted that he supplied 109 tons of sugar cane in 1951. At a rate of $7.24 per ton he would have received a gross return of $789.16.

Therefore, while the increased share did not mean additional returns for many *metayers*, it did allow the more successful among them to derive substantive returns from their investments. It must also be noted that the volume of wealth accumulated by the *metayer* was not amenable to official accounting, that is to say, all of the *metayer's* sources of income were not accounted for in the official statistics.

The contributor's earnings did not constitute his aggregate revenue from agriculture, because many small farmers were shifting their investment from sugar to bananas.[69] Hence, to many, the income from sugar was only a part of their total revenue. A survey conducted by a team of experts confirmed this argument. According to them, "the people seem to be content in living and leading a meager and abject life [but] hundreds of dollars were spent to buy the best dresses and shoes for the occasion and to have a good time". Bewildered by the findings, they lamented:

> One naturally asks the question: *Where does the money come from* and can so much be spent in the light of seemingly poor conditions (emphasis added)?[70]

The team of experts fell prey to the often repeated problem of accounting peasant wealth, as well as determining their world view of human and physical development. A peasant/landlord in Dennery perceived the situation differently, and noted that "a lot of the people have *big savings* . . . The existence of wooden houses is a direct result of their lack of access to land . . . Anyone who has a house spot, it can be certain will build a modern building. It shows the capability is always there."[71]

In a monetary economy a peasant has the alternative of saving in the form of cash or goods. The holding of certain goods may stabilize the value of his money, and may be preferred insofar as it can be a direct contribution to the feasts of neighbours and kinsfolk, or it can be easily converted into cash for festivals. In addition, it offers instant liquidity for such occasions. Money may be converted into such forms as jewellery, animals, and such non-perishables, which are in great demand or can serve as easily realizable assets [Firth and Yamey 1969: 23-24]. This argument is theoretically and empirically relevant, since the standard approach utilized in determining and accounting wealth, savings, and investment is limited when analysing peasant accumulation.

Table 6.4: Cane Production by Producers, 1951

Producer	Canes	Canes/Sugar	Sugar
Roseau			
Co. Estate	39,321	10.11	3,890
Contributors	1,279	10.11	122
Metayers	2,636	10.11	422
	43,236	10.11	4,434
Cul-de-Sac			
Co. Estate	27,498	10.20	2,695
Contributors	2,336	10.20	229
Metayers	2,645	10.20	259
	32,479	10.20	3,183
Dennery			
Co. Estate	14,218	10.24	1,389
Contributors	5,665	10.24	553
Metayers	2,687	10.24	262
	22,570	10.24	2,204

Source: *Report of the Commission to Enquire into the Stoppage of Work at the Sugar Factories in St Lucia in March 1952 and into the Adequacy of the Existing Wage-Fixing Machinery in that Colony*, p.2.

Peasants in St Lucia have never altogether resorted to the official or formal financial institutions in their day-to-day monetary transactions. Based on their own precautionary, speculative, and transactional demand for money, they have always utilized a variety of savings and investment mechanisms for their own advantage. In 1845 a Colonial Office Report noted that the labouring class kept "an abundantly stocked yard of pigs, goats and poultry".[72] In addition, they utilized the facilities of the Friendly Societies. There are also reports of rural farmers burying their monies in holes.

Labourers, *metayers*, and small and middle proprietors and labourers in Dennery, as we have noted, have always kept an abundance of domesticated animals, which they easily convert into liquid assets when the need arises. In fact, the district has always been noted for the prevalence of stray animals. These animals are sold and slaughtered on feast days and special holidays. Therefore, it is quite possible that a researcher may impute his concept of poverty and development to a particular rural population which is not necessarily congruent with their world view.

Table 6.4 indicates that sugar yields were generally low. The quantity of sugar derived from 81,037 tons of estate canes was 7,974 tons, and the yield from 17,248 tons of canes produced by contributors and *metayers* was only 1,847 tons. On the average, sugar yields were about 10 percent of total tonnage harvested. There are several factors that contribute to low yields, including cane type, soil type and technological application. Thomas [1988: 27] noted that, up to World War II, the estates depended on increased acreage to raise production levels while "application of technology" consumed only a small part of their capital investment. Estate yields dropped from 3.91 tons in 1950, to 3.42 tons in 1954.[73] The collapse of the large plantations was manifest in their increasing dependence on peasant production. It appears that the relationship between the peasantry and the plantations was assuming a zero sum character.

Sugar production was finally replaced by bananas in 1963 (the issue is discussed in the following section) but, even under more modern techniques of production, the *metayer* continued to cultivate very successfully in the banana industry. The sharecropper had always attempted to take advantage of the opportunities for accumulation that the system offered him. A 50-year-old *metayer* at Errad in the highland area, said that he cultivated 6.5 acres of bananas and ground provisions on a 50:50 share on receipt less labour cost. The coconuts on the land were also reaped and processed into copra on a 50:50 share.

A small proprietor/landlord who was cultivating 10 acres on the sharecropping system informed the interviewer that his arrangement with the tenant evolved from non-formal to formal agreements. In 1981 the sharecropper received 80 percent of the net monthly income. In that year the tenant produced between 140 and 160 cartons of bananas fortnightly (a carton holds 30 lb of bananas and the average producer price in 1981 was 16.21 cents per lb). The cultivator kept whatever income was obtained from the sale of coconuts and citrus products, in exchange for caring for the fruit trees in addition to the bananas. In 1983 they signed a contract that provided for a 75 percent share of the gross income for the sharecropper. By 1989 the *metayer's* savings had allowed him to own a van which he bought on hire purchase in 1986 at a total cost of EC$22,000, and begin the construction of a $150,000 house, offering liquid asset for collateral. "A number of *metayers* at Gadette [in the lowland area] purchased lands from . . . Dennery Farm Company . . . with liquid cash."[74]

A reliable informant who had access to some of the *metayers'* bank accounts offered evidence relating to the level of accumulation of those tenants. According to her, "they are presently the big earners in the region making about EC$2,000.00 fortnightly from sale of bananas in gross income. Some earn in the range of $500.00 to $2,000.00 a fortnight . . . In the Gadette area, they own 10 vehicles costing over EC$ 40,000.00 each". Of the 53 fields investigated at Tete Chemin in 1984, Sahr found that there were seven comprising 19.75 acres of land cultivated by sharecroppers [Sahr 1987: 63].

It would be interesting to trace the development of the marginal class of sharecroppers so as to analyse its dynamics under commercial agriculture. This is a theoretically relevant issue, insofar as the Marxists contend that sharecropping is ultimately destroyed with the development of capitalism in agriculture [Pearce 1983]. Although this is an area for further research, the two case studies strongly suggest that the sharecroppers possess tremendous potential for substantial accumulation of wealth.

The utilization of sharecropping under conditions of advanced productive techniques underscores the utility in the Marxist argument that sharecropping can be used even under conditions of capitalist penetration in agriculture [Byres 1983: 10].

Increasing Control of the Medium Peasants

The medium farmers were the major benefactors or accumulators in export agriculture in the post–World War II period. What is referred to in Table 6.5 as 'Farmers Canes' constituted the produce of the middle peasants who sold to the estates at the current market price (the produce of *metayers* and contributors were aggregated with that of the estates). The production level for 1940-1943 compared with that of 1941-1942 suggests a contraction in sugar cane production between 1940 and 1943. The declining trend was evident on both estates and farms, and thus reinforces the fact that the sugar industry was in decline.

Table 6.5: Sugar Production in Dennery, Cul-de-Sac, and Roseau, 1940-1943

Area		1941-1942			1940-1943	
	Estate Canes	Farmers Canes	Total Canes	Estate Canes	Farmers Canes	Total
Dennery	11,398	5,018	16,416	—	—	11,117
Cul-de-Sac	24,075	4,973	29,048	18,043	3,494	21,537
Roseau	29,814	3,295	33,109	27,262	2,513	29,775

Source: "The sugar industry in the Caribbean", Caribbean Commission Crop Inquiry Series, No. 6, p.178.

Some middle peasants entered into a variety of sharecropping agreements, mainly verbal contracts, with their kinsmen or neighbouring labourers, who sought permission to occupy idle lands. A number of those tenants have ultimately become owners of those lands which they occupied by sufferance, that is, gaining legal rights to the lands after more than 30 years of continuous occupation in the absence of legal documents proving their status as tenants.[75] Significant examples are the Sargusingh family in the highland area, who operated part of their lands on shares, and Arnold Nicholas in the lowland area, who operated more than 600 acres of land at La Resource and Gadette on shares. The practice on Nicholas' land, which was continued by the other generations, was brought to an end by a court decision in 1989.

The operative conditions under *metayage* on the Nicholas lands varied with each *metayer*. Generally the share of the landlord was determined by the proximity of the land to the tenant, the development of the land at the time of acquisition, and the capital input of the landlord other than the land. A tenant working on lands with ratoons was expected to surrender three-fourths of his produce; occupiers of undeveloped land returned one-half of their proceeds; and where advances were made the landlord expected three-fourths of the produce. A member of the Nicholas family perceived the benefit to be derived from share cultivation as twofold. He explained that, in addition to his prescribed share, the landlord received ground provisions on feast days and holidays such as Christmas, New Year, and Easter. Where the share was paid in cash, *metayers* were expected to produce the receipt of the sale. In the half-share arrangement, where the *metayer* cultivated bananas, the consensus was to alternate the harvest between landlord and *metayer*. According to him, hundreds of acres of land which would have otherwise remained fallow appreciated in value under the system. This was done at very little cost to the landlord. The land is now worth about ten times its value ten years ago, and is presently valued by the commercial banks as first class primary agricultural land. It can be easily mortgaged or used as collateral, since it is not partitioned.

In the postwar period, while plantation production was steadily declining, peasant production was rapidly increasing. Small farmers (those delivering 0-5 tons) and the medium farmers (those delivering 6-50 tons) were gradually gaining control of the agricultural sector. Large farmers (those delivering 51-500 tons) were gradually losing their monopoly in export agriculture. Table 6.6 demonstrates the degree of peasant involvement in export agriculture following the fall of the planter class.

Table 6.6: Cane Farmers' Production by Size, 1950-1954

Crop	0-5	6-50	51-500	Total	Tonnage
1950	560	668	62	1,290	19,435
1951	770	708	55	1,533	18,879
1952	868	792	46	1,706	18,672
1953	989	854	49	1,892	18,197
1954	1,019	767	48	1,834	19,288

Source: British West Indies Sugar Association (Inc.), p.99.

The number of peasant farmers involved in cane production increased over the period by 544 farmers, that is, a rise of 42.17 percent. The number of small farmers rose by 81.96 percent (459), while the number of middle farmers increased by 14.82 percent (99).

The sharp decline in the involvement of the class of medium farmers in the sugar industry was due primarily to a shift in investment from sugar to banana cultivation. It was certainly more difficult for the small farmers (mainly contributors) to diversify, insofar as they were obligated and indebted to the estates. Of the estimated 2,500 acres of bananas under cultivation, peasant production accounted for some "1,500 acres and estate production for 1,000 acres" [Colonial Office Report 1955-6: 31]. There were 394 registered members in the Banana Growers Association.

On the contrary, the number of large farmers in production dropped by 22.58 percent to 48. Although the data presented are not sufficient to allow for any generalization, it is evident that the peasantry was gradually gaining control of the sugar industry. However, the prospects in the industry were not promising, as average yields were falling. The yields dropped by 0.76 percent (147 tons) for the period. Table 6.7 traces changes in aggregate earnings in the banana industry.

Despite the sharp decline in 1952, the data show a general and persistent rise in the gross aggregate annual income generated by cane farmers in the sugar industry. Aggregate income increased by 38.94 percent in four years. However, this is not an indicator of profitability, as the industry was in a state of collapse. Although the average acreage under cultivation on the estates increased from 726 in 1951 to 826 in 1954 (13.77 percent), the yield per acre decreased from 3.91 tons per acre in 1950 to 3.55 tons per acre in 1953.[76]

Table 6.7: Cane Prices and Gross Earnings, 1950-1954 (Long Tons $)

Year	Price	Gross Earnings
1950	$7.20	$139,932.00
1951	7.60	143,480.00
1952	8.80	104,313.60
1953	10.08	183,525.76
1954	10.08	194,423.04

Source: *British West Indies Sugar Association (Inc.)*, p. 93. Computed from Tables 6.5 and 6.6.

The capital/labour struggles in the early part of 1953 so strained the capital/labour relation in the sugar industry that the planters who owned the three valley limited companies informed the colonial administration of their intention to go into voluntary liquidation on 7 March 1953 [Helminiak 1972: 42]. In that same year all three sugar companies (Dennery Factory Company Ltd, Cul-de-Sac Factory Company Ltd, and Roseau Factory Company Ltd) were taken over by a public company, Sugar Manufacturers Limited. The European capitalists' control of the sugar industry had come to an end.

The new management of the estates could not salvage the plantations from their inevitable collapse, and the government initiated a programme to assist the industry. An agricultural association, nurseries, and a credit scheme were established to provide technical assistance. A revolving fund of EC$24,000 was provided under a Colonial Development and Welfare Scheme for the Credit Scheme, but this could not keep the plantations in production at a time of stiff competition from bananas and persistent labour/capital conflict.[77]

The influence and increasing control of the peasantry in export agriculture in the postwar period was facilitated by the fall of the planter class. As the most financially able agrarian class, it became very influential in the ailing industry. Table 6.8 shows the leading agricultural exports in the terminal phase of the sugar industry. There is evidence of an interrupted increase in banana production and a marginal increment in sugar output. While the value of sugar appreciated at a moderate rate, the value of bananas increased at a faster rate between 1954 and 1956.

In comparison to the moderate increase in the value of sugar exported between 1955 and 1956, the value of bananas exported increased phenomenally in the same period. By 1956, bananas had become almost as important as sugar as a foreign exchange earner. In 1956 "an estimated 2,500 acres" of bananas were under cultivation, "peasant production covering some 1,500 acres".[78] Three years after the European proprietors went into liquidation, the peasantry had become the more economically powerful in the ailing industry.

Table 6.8: St Lucia's Principal Exports, 1954-1956

Year	SUGAR Quantity (Tons)	SUGAR Value (£)	BANANAS Quantity (Tons)	BANANAS Value (£)
1954	7,301	1,309,899	306,432	506,421
1955	9,053	1,559,090	445,627	642,778
1956	9,125	1,598,687	753,950	1,537,220

Source: Land Settlement Scheme for St Lucia Based on a Survey of the Agricultural and Social Conditions of the Island, 1958, p.34.

While the sugar industry was collapsing, a number of external commercial and economic forces were creating the conditions for an agricultural change in the colony. In the postwar period an agricultural change was effected in the agrarian economy—a shift from sugar to bananas. The factors that facilitated this change were as follows:

1. An increasing demand for bananas in the United Kingdom in the 1950s
2. In 1954 Geest Industries (WI) Ltd replaced Antilles Products with the contract of purchasing "all marketable bananas produced without upper limit", thus providing unlimited marketable facilities

3. In 1956, the United Kingdom expanded protection for Commonwealth bananas from £2 10s to £7 10s per ton and placed limitation on banana imports from low cost Latin American producers through the establishment of a dollar area import quota of 4,000 tons. Therefore, in 1957 when the valley strike dealt the fatal blow to the sugar industry, it was simply the culmination of a fluctuating but steady collapse.[79]

A Colonial Office Report in 1957 stated: "The cane farmers' contribution to sugar production fell from 11,317 tons in 1957 to 10,317 tons in 1958, as a result of conversion to bananas."[80] In 1964 "the production of sugar ceased with the conversion of the last sugar producing valley to banana cultivation".[81]

Peasant Influence in Politics

The increasing economic control of the peasantry allowed them a greater role in the administration of the colony. With increasing acquisition of wealth, many of the middle peasants became eligible for candidacy in electoral politics, in spite of the narrow franchise. The elective franchise also became more accessible to other small peasants who were previously debarred from voting. In 1937, 1,509 electors, compared with 366 in 1928, voted five candidates into public office for the four constituencies.[82] The number of eligible electorates on the same narrow franchise had increased by 1,143 in less than 10 years, more than three times the number of voters in 1928. This shows a phenomenal increase in the political participation of the peasantry.

When adult suffrage was granted in 1951, the peasants did not play an active role in national politics, preferring to make their contributions at the local level. Statistics for local authority elections between 1953 and 1958 reveal that planters and proprietors won 13 of the 50 seats contested (26 percent). Farmers exercised very strong influence in the agricultural belts. For example, the Micoud local authority was dominated by farmers between 1953 and 1958 and, in most cases, they were elected unopposed.[83]

There are several factors that account for the low participation of the peasantry in national politics. Peasant farmers demand more from the political and the labour movement than the agro-proletarians. Their demands transcend the alleviation of the problems of low wages, poor working conditions, and unemployment. Their set of priorities include more than better lands, increased prices for their products, more insurance coverage, access to and ability to purchase chemicals, adequate transportation and feeder roads, among other things. The demands of colonial politics were better suited to the style, skills, and tactics of the politicians and labour leaders. Mass demonstrations, the rhetoric of protest, wage bargaining are not their strengths [Emmanuel 1981: 23]. The more complex farming community calls for less drama and more sustained national application of bureaucratic and fiscal resources of the state. Migdad [1977: 21-22] contends that peasants shun "participation in unstable societal institutions in which they could suffer from the exploitation of those above them on the social ladder". Their practical goals are limited and are not "directed to the policy level of politics".

Summary

This final chapter of analysis is characterized by some of the analytical defects inherent in the previous ones. While the appraisal is deficient in data presentation, the quantitative evidence presented is fairly representative of the social classes discussed. The analysis would have been enhanced if statistical trends were accessible for all the variables discussed, but the historical evidence presented reinforces the issues raised.

The local economy changed significantly between 1929 and 1957. By the end of 1957 peasant production had replaced plantation production. The capitalist planters lost their control over the agricultural sector and the direction of the economy, and were replaced by a peasant class, which dominated export agriculture.

By the end of the third phase of *metayage* (1929-1957), a 'revolution' had been effected in the underdeveloped capitalist agrarian economy. In the immediate post–World War II period, peasant involvement in export agriculture was radically altered. Initially, the estates became very dependent on peasant production. The medium peasants enjoyed an increasing control of the sugar industry, in spite of the high production costs consequent on low yields. *Metayers* obtained a greater proportion of the share of the produce (four-fifths, as compared to two-thirds in the previous phase).

Even before the shift from sugar to bananas in 1957 the peasantry had already established itself as the dominant primary producer. In 1953 the European capitalists who owned and controlled the sugar production on the three valleys went into liquidation. The estates were taken over by public companies.

The banana 'revolution' brought about the replacement of an estate crop by a smallholding crop and, thus, radically altered the character of accumulation in the agrarian economy. Under banana cultivation a small farmer is able to realize a profit from his investment, given his access to labour and the product market. The large and medium farmers, however, generated a greater return from their investment, given their greater access to capital, labour, and technology. The 1973 Agricultural Census noted that "small peasants owning less than five acres of land own 82.1 percent of all farms. Middle strata peasants owning land between five and 50 acres own 16.4 percent of all farms, while the large landowners form only one percent of farmers . . . Over 75 percent of banana farmers have 10 acres or less" [Latin American Bureau 1987: 67, 68].

Every banana farmer plays an active role in the global capitalist system. He occupies a specific place in the production process, and assumes a specific role in the organization of labour. However, the mode of acquiring wealth, as well as the nature of ownership of the means of production, varies. The few *sharecroppers* remain dependent on the 'capitalist/peasant/landlord' (by then, the plantations no longer practised *metayage*), exert very little influence on the labour market—mainly utilizing family labour—and depend on their share of the produce for sustenance. This is not to say that a landlord class developed, because after the shift to banana cultivation the *metayer* class was marginalized. The medium peasants expanded their scale of production to take advantage of increasing demand. They constituted an *intermediary class* exhibiting some characteristics of Lenin's small peasant, holding a small plot through tenancy and being unable to hire outside labour [Lenin 1980].

The *small peasant* owns his means of production (he occupies and works very few acres of land and has a small capital outlay). Insofar as he utilizes both family labour and wage labour, his labour use is greater than that of the sharecropper, but smaller than that of the middle peasant.

There are indeed some parallels between the small peasant described here and the middle peasant in Russia discussed by Lenin [1980]. They both own their lands and generate just more than what is required for their family to subsist. This small surplus is reinvested for profit. Relative to the Leninist capitalist peasant, the small peasant is, however, more adequately described as 'capitalist/peasant', that is to say, he owns 'peasant means of production' but maintains 'capitalist relations of production'. This social class has a greater or lesser degree of involvement in the labour, money, and product markets.

The *medium peasants* compete with the plantation proprietors on the labour market, and earn large profits from their investments. They exert the most political influence on the Banana Growers Association (the estates and Geest Industries Ltd have no membership in the Association). Up to 1980 only those growers who sold 500 lb of bananas or more per week could be elected as representatives of the 40 districts, or present themselves for election to the board of directors. In spite of the fact that this group comprises only 16.73 percent of the total number of farmers in the industry (small farmers comprise 82.1 percent), it exercises considerable political and economic influence in an economy where bananas are the dominant foreign exchange earner [Latin America Bureau 1987: 60-70].

This social class, however, is different from the European 'capitalist/ peasant class' which is discussed by both Marx or Lenin. Although as a 'class', it is deeply integrated into the capitalist market and employs a number of wage labourers, it retains a peasant outlook, insofar as it is involved in direct production. In addition, few of the farmers have access to the capital, land, and technology that the commercial farmers enjoy, and most do not utilize the same level of rational accounting and organization of production as the capitalists who own and manage the plantations. As a class, they are more precapitalist in their approach to production and labour use. In addition, their labour use is different from the 'capitalist' who appropriates surplus labour from the 'agro-proletarian' on the plantations [Emmanuel 1981].

CHAPTER SEVEN

Conclusion

Sharecropping has never had overwhelming success anywhere it has been practised. It was moderately successful in the Gezira Development Scheme in Sudan, formalized between 1925 and the mid twentieth century. The scheme was efficiently managed and was characterized by a high level of administrative accountability as well as a strong solidarity among tenants [Gaitskell 1959]. "The only instance that occurs in colonial history of the adoption of the Metairie or Metayer system, is that of Prince Edwards Island" [Breen 1844: 303].

The establishment, persistence, and decline of *metayage* in export agriculture was conditioned by exogenous and endogenous factors. Between 1846 and 1957, the plantation system as part of the global capitalist system was forced to respond to many adverse developments within the world economy. Commercial, economic, and political shocks (such as the Sugar Duties Act of 1846), protectionist policy for beet sugar, and impact of war dislocations, among others, created conditions favourable for the practice of *metayage*.

It was factors in the domestic political economy most of all that facilitated the stability of *metayage*. The cycles of crises and depression caused by shortages of capital, declining productivity levels, labour shortages, application of backward technology, and mismanagement created conditions for more than one hundred years of *metayage* cultivation. However, it was the established land culture which facilitated the acceptance and establishment of the tenure type. The strong sentimental attachment to land, the practice of communal ownership of land, labour sharing, product sharing, and a general hunger for land and independence were all compatible with the structural arrangement of *metayage*. The system of tenure, therefore, was suited to the orientation of the rural population. This also explains its survival, though on a much smaller scale.

Some Historical and Theoretical Issues

The conventional wisdom portrays *metayage* as a temporary expedient, and *metayers* as mere victims of an experiment. The experience of St Lucia, where the system was most extensively practised, however, reveals the reverse. To say that *metayers* were merely victims is far from being accurate, insofar as they were both victims and victors. The *metayage* system altered the very character of the social structure in St Lucia. Notwithstanding the many *metayers* who were exploited and pauperized, the system produced a new class of propertied freeholders, who continue to influence the economic and political directions of the society in a significant way.

Both planters and *metayers* were individualistic, each seeking to maximize their advantage. The planters sought to appropriate surplus through labour use; the *metayers* sought to take advantage of weaknesses inherent in the arrangement, and the opportunities to accumulate wealth. In periods of crisis and depression the planters and the *metayers* were drawn together, but in times of boom and upswings they were separated. But oftentimes they were forced to succumb to the exigencies of the economic conditions, compelled to be satisfied with such returns sufficient to allow them to remain in production. In some periods the *metayers* received just enough to sustain themselves and their families and were driven by a determination to operate by their own timetables. In some cycles the planters were faced with two alternatives: to opt for *metayage* or to liquidate. Those who favoured sharecropping were prepared to take their losses in the short run in anticipation of a long-run windfall.

In a theoretical sense, the *metayers,* as a group, constituted an agrarian class in an objective and a subjective sense. They had a specific relation to the means of production, participated in the social organization of labour, and acquired their wealth through share tenancy. They developed a solidarity (a consciousness) among themselves in their struggle to present a united front against the planters (their enemy). The *coud de main* (helping hand) was their main strategy for cutting labour costs.

Among the rural classes the *metayer* class exerted the most influence on rural life and rural transformation in the immediate postemancipation period (1840-1900). The records strongly suggest that any other agrarian class in export agriculture (the dominant source of income generation) would have been very marginal in the immediate postemancipation period, until the turn of the century. The colonial records do not make reference to a free peasantry in capitalist production until 1897. The small proprietors, which the literature refers to as the free peasantry at the beginning of the twentieth century, were formerly *metayers*. Therefore, we cannot speak of the contribution of a free peasantry in St Lucia between 1838 and 1957 without running the risk of losing significant analytical details.

The peripheral or insignificant role attributed to *metayers* in rural transformation in St Lucia is a consequence of defective and ahistorical analyses under the guise of history. This criticism is only applicable to specific micro level analyses. Such analytical inadequacy is likely to occur if one attempts to treat the peasantry as if it were a homogeneous class—it is heterogeneous and different in character in each specific context.

If one omits the analytical significance of the *metayer* as a specific agrarian social class in preference to a discussion of a heterogeneous peasantry in St Lucia, one misses crucial dynamics of the development of that particular peasantry. The *metayer* class was the 'feeder class' for what now constitutes the free peasantry in export agriculture. Therefore, it is only by tracing the growth of the peasantry from *metayage* that we are able to grasp the dynamics of change.

What was the mechanism of change that accelerated the evolution of the peasant classes? The medium of accumulation determined the differentiation of the *metayer* class and dictated the pace of integration of the peasantry into capitalism. As the metayers accumulated, they acquired lands and became smallholders. Smallholders who lacked the working capital became contributors, selling their produce at prearranged prices to the plantations in exchange for advances received; larger *metayers* became independent proprietors and sold their produce to the plantations at the current market rates.

As the peasant classes became more and more involved in the capitalist market their economic orientations were altered. One cannot operate successfully in a system without adapting to the rules of that system. Therefore, the emergent peasant classes had to operate within the capitalist system and utilize capitalist mechanisms if they were to maximize profits.

By the end of the period two 'capitalist' classes had emerged among the labouring class. A class of small farmers which is 'capitalist/peasant' in character and a class of medium farmers which is 'capitalist' in behaviour.

However, the agrarian classes are not prototypes, insofar as they operate within a relatively underdeveloped capitalist economy. In addition, many of their productive activities are still precapitalist in nature.

From the foregoing discussions we can conclude that it would be inaccurate to say that *metayage* practice in St Lucia was a failure. The success or failure of the practice must be seen in terms of its overall impact on the society, that is, how the practice affected the social, political and economic systems, as well as the physical infrastructural arrangements of the rural communities. Perhaps an appropriate approach is to audit the system in an accounting sense. On the debit side, the antagonist can contend that, inasmuch as the planters did succeed in utilizing the underdeveloped capitalist system to exploit the majority of *metayers*, it was a failure. On the credit side, the protagonist can argue that insofar as the system altered the social structure of the colony and, more particularly, the social structure of the rural communities where it was practised, it was, therefore, a success. *Metayage*, directly or indirectly,

- altered the property relation in the colony, leaving behind new and distinct agrarian social classes
- orchestrated the decline and fall of the plantocracy, thus transferring the export agricultural sector to the new indigenous small proprietorship
- alleviated the social and economic conditions of the labouring class in St Lucia, thus allowing the victims of capitalism, slavery, and racism, through continuous struggle to obtain a measure of redress

In this regard, the credits appear to outweigh the debits. The ultimate effects of the system on the social structure illustrates, therefore, a perpetual success that renders its failures secondary, if not peripheral. One may even argue that its victims are exalted through its victors.

The study does not argue that *metayage* was the sole mechanism of change, but that it was the primary mechanism of change. Although the analysis in chapters 4 through 6 does not provide sufficient quantifiable proof, it does trace the evolution and development of an agrarian capitalist class in St Lucia from 1840 to 1957, and therefore supports our hypothesis.

Table 7.1 summarizes the development of the agrarian 'capitalist' classes during the period of *metayage*. In each phase, the *metayer's* share increased from one-half to four-fifths of the produce, thus aiding some former labourers to climb higher up the socioeconomic ladder, enjoy a greater degree of independence, and become more integrated into the global capitalist system.

Table 7.1: The Emergence of a 'Capitalist' Class under *Metayage*: a Summary

	Phase I (1840-1897)	Phase II (1897-1929)	Phase III (1929-1957)
Type of cultivators	*Metayers*	*Metayers* Contributors Proprietors	*Metayers* Contributor Proprietors
Metayer's share of produce	One-half	Two-thirds	Four-fifths
Change in property relations	Labourers Tenants	Labourers Tenants Freeholders	Labourers Tenants Freeholders Landlords
Change in market relations	Labour	Labour Product	Labour Product Capital
Change in involvement in foreign market	None	Minimum	Maximum
Power relations	Little	Some	More
Social solidarity	Informal	Informal	Informal Formal

The Impact of *Metayage* on the Social Structure

A reconstituted social structure developed from the *metayer* class, the 'feeder class'. Prior to the introduction of the *metayage* system, the social actors in export agriculture were the planters (the 'capitalist class') and the labourers (the 'agro-proletariat'), but by 1957 the actors comprised a free, differentiated peasantry, which dominated production and accumulation.

Table 7.2 derives a typology of the peasant classes presently involved in export agriculture. Notwithstanding their similarities, the classes are clearly distinct with respect to their relation to means of production, the nature of the forces of production, labour use, and the mode of acquiring wealth. Three classes of peasant farmers are presently involved in the production process—*metayers*, small farmers, and medium farmers.

A process of differentiation corresponding to that discussed by Lenin [1980] in relation to the decomposition of feudalism in Russia is evident. The *metayers*, like Lenin's small peasants, cultivate as share tenants, utilize mainly family labour, have very little access to capital, and derive their wealth primarily from their share of the produce. The small farmers

TABLE 7.2: A TYPOLOGY OF PEASANT CLASSES PRESENTLY ENGAGED IN EXPORT AGRICULTURE IN ST LUCIA

	SEMI-PROLE.*	*METAYERS*	SMALL FARMERS (0–5)**	MEDIUM FARMERS (6–50)**
Access to land	Squatting Family land Permission	Tenancy	Freehold	Freehold
Access to capital	None	None	Little capital outlay	Large capital outlay
Access to labour	Wage earner	Direct producer Family labour	Direct producer Family labour Small gang of labourers	Direct producer Large gang of labourers
Crop produced	Provisions	Bananas	Bananas	Bananas
Technology	Traditional	Traditional	Modern	Modern
Mode of acquiring wealth	Wages	Shares	Profits	Profits

* semi-proletarians
** figures in acres

(formerly contributors) own only a few acres of land (0-5 acres), have limited working capital, and exert little influence on the labour market, but generate sufficient income to allow for reinvestment. This class manifests characteristics of Lenin's middle capitalist peasant, but utilizes a combination of capitalist and precapitalist practices, as well as distinct customary forms of labour organization. The medium farmers (formerly proprietors) own several acres of land and operate as 'capitalist' entrepreneurs. They have access to capital (working capital, credit facilities, and technology), and compete with the plantation for labour. They acquire their share of wealth primarily through profits generated through agricultural production, and operate in many ways like Lenin's big capitalist peasant class.

The term 'capitalist' refers neither to the industrial capitalist discussed by Marx nor the bourgeois capitalist peasant discussed by Lenin. The concept 'capitalist' is used primarily to refer to the involvement and integration of the peasant classes into the capitalist system and their utilization of capitalist logics of accumulation under capitalist and precapitalist arrangements.

Some Policy Implications

What implications does this work have for development administrators and planners? The study reinforces two things: the relevance of history to an understanding of change and policy prescriptions, and that development and change are best understood within the dynamics of production and circulation

Planning and Development Strategy

There is need for a consortium approach to development. Sociologists, economists, administrators, and physical planners must come in the planning process to provide a holistic development strategy. The sociologist is required for the "transformation of problems by introducing concepts which refer to variables overlooked in the common sense view of the policy maker. At times, the concept leads to a statement of the problem that is diametrically opposed to that of the policy maker" [Merton 1949: 178].

It is evident that the agrarian 'capitalist' classes presently involved in export agriculture have substantive potential for growth and development. This dynamism requires the tailoring of programmes and the establishment of facilities to enhance their increasing capacity to contribute to gross national product and foreign exchange generation. Access to credit facilities will enhance their ability to take advantage of a changing capitalist market into which they are integrated.

The operation of the 'capitalist' peasants points to their desire for economic efficiency and, therefore, demands increasing access to modern techniques and methods of farming. Given their sentimental attachment to the land, their use of distinct precapitalist and customary economic practices, such as their approach to labour use and labour organization, state agrarian practices and policies must not be designed so as to appear antagonistic to their traditional approaches to agriculture. The *metayers* and small farmers seek to maximize profits based on their understanding of the economic laws which govern production and, as a result, their approach will not always be compatible with that of the state agency. They must not be made to think that state agricultural policies are imposed upon them from above. The 'capitalist' peasant classes, on the other hand, utilize more rational economic approaches and will most likely be more receptive to changes and formal economic approaches.

Attempts to integrate the peasantry into state farms are most likely to be more successful among landless peasants who are eager to grasp such opportunities for independence; freeholders aspire to preserve the right to establish their own rhythm.

The preceding discussions point to a particular dominant rural culture which accelerated the emergence of a free capitalist peasantry in export agriculture and aided the development of colony. This culture reflected a communal approach to production, among other factors, that facilitated the social and structural differentiation that was peculiar to the rural districts. The practice of land sharing, product sharing, and labour sharing was the sociocultural mechanism that empowered the process. This suggests that inherent in the rural society is a social solidarity, among other things, that invested and aided those successful *metayers* with support mechanisms, and facilitated their potential to accumulate even under asymmetric conditions.

This spirit of cooperation and community, which prevails in the rural districts, offers a viable potential for the organization of labour for agricultural, infrastructural, and community development. The prevalence of the rural work culture proposes the utilization of the self-help and cooperative approach to development.

The cooperative approach to development harmonizes perfectly with communities with strong consanguinity, as is the case of most communities in St Lucia. The mutual utilization of skills, labour, organizational talents, and managerial expertise for self and community improvement is not alien to a population that has utilized similar mechanisms to carve out a place for itself.

APPENDIX A

Copy of a Metayage Contract Made on the 31st of July 1848, Between Cultivators and Proprietor of Black Bay Estate in Vieux Fort, the Fourth Judical District in the Colony of St Lucia

1. The Association by Notorial Act dated 31st July 1848 to be enforced during 3 years with stipulations as follows:
2. To cultivate and keep in order the Black Bay Estate, the Proprietor furnishing at his expense and in good condition, the Sugar Houses, Bagass Houses, Coppers, Boilers, and other utensils necessary for the manufacturing process.
   Also a new Water Mill for the second crop, and should the one at present in use be injured, to be repaired at the cost of the Proprietor alone.
3. The following animals included in the Association viz. eleven mules in good order, twelve draft oxen, five cows, a calf, and a bull, and should any die, or become unserviceable during the Association, they are to be replaced, each party paying an equal portion of the cost.
4. All acquisition of animals to be only the parties, share and share alike, and at the expiration of the contract the increase, if any, to be shared in like manner.
5. The expense of Manufacturing the Sugar, Rum and Molasses, to be borne in common as well as of the Caulikeeper, keeping in repair the Buildings and Manufacturies. The manager's salary not to exceed £60 p/an, to pay all Colonial duties commencing from 1st January 1849, also to furnish the necessary utensils for containing the Rum, Sugar, and Molasses and to keep in refrain the Canal in Mason Work or Planks.

6. The cultivators are obliged to perform the following work:
   To cultivate the Estate, dig the holes, plant the canes, furnish and deposit manure as often as necessary, weed the plant canes at least 4 times and Ratoons not less than 3 times yearly.
7. The cultivators are also bound to keep the Canal and water course to the mill in order, supply fuel, keep the pastures clean, the road and fence in repair, carry manure to the field, furnish the cattle pans, with trash, carry lumber and materials for the construction of the new Mill and repairing the buildings, and in default of so doing the same to be done at the expense of the defaulter.
8. The canes now on the Estate, whether plants or ratoons, having received a first weeding and been kept in order at the sole expense of the proprietor, to be cultivated and manufactured as the others with the exception that the cultivators receive a third instead of half the produce.
9. All Sugar, Rum and Molasses, manufactured from and after the 1st January 1849 after being put into casks, and the colonial duty and other expenses paid by the association, to be separated or divided in the Sugar House in the following manner viz. Half of the produce to the Proprietor and the other half to the cultivators, each individual disposing of his share or portion in any manner he may think proper.
10. The cultivators are obliged on being paid for the same to transport the produce accruing to the proprietor in carts and with the cattle of the Estate, to the place of embarkation, and the proprietor cannot on his part prevent the cultivators from using the same mode of conveyance for their produce, which they are allowed to place in security free of expense in the magazine of the proprietor on the Bay where they are to be shipped.
11. The cultivators to choose their provision grounds in such places as may not be eligible for canes, and they are allowed to put cattle being their property with the herd.
12. If any cultivator neglect to care his portion of canes, a verbal notice to be given him by the Manager 'in presence of witnesses' to clean it within 8 days, and in default of so doing on application to competent authority, the said cultivator can be compelled to abandon the same for the benefit of the society.
13. At the dissolution of the Society, the association will divide in equal portions, the surplus of cattle, should there be any, and on the other-hand the cultivators are obliged to make good any deficiency that may occur.
14. No cultivator can transfer his claim to any person without the formal consent of the proprietor.
15. All persons cultivating provision grounds on the Estate, and are not parties to the present contract, to give a day in each week to the Society, and no Garden is to exceed 3 acres without the consent of the proprietor.
16. The agreement can be renewed for 6 years should the parties be willing to do so. Three months notice to be given to that effect.
17. The cost of the Notorial Act to be paid by the Proprietor and Cultivator.

Source: C.O. 258/16 J. Johnson to V. Drysdale, Stipendiary Magistrate Report for the 4th District Dec. 31, 1848, pp.432-33.

APPENDIX B

Sections 886 and 887 of the Code of Civil Procedure: St Lucia 1950

886. In cases of leases of land for agricultural purposes, when there is no written lease fixing the day upon which the lease is to be terminated, and in the cases in which land is held by permission of the proprietor without lease, the plaintiff must serve upon the defendant a notice to quit and deliver possession on the 1st day of August then ensuing. If the defendant has no known residence in the Colony, the bailiff posts on the dwelling house or other conspicuous place on the property leased, having previously notified one of the neighbours of his intention to do so, a notice to quit and deliver possession. The notice must be served or posted as the case may be one month at least previous to the institution of an action to annul the lease or to give up possession. The lessor may sue by the same action for the rent due him.

887. If not otherwise agreed upon by the parties, the court may grant compensation for the growing crop to a defendant ordered to give up possession to the owner or lessor, unless the owner or lessor allow the occupier or tenant to take away such crop, in which last case the court gives a reasonable time for its removal.

Source: *Caribbean Economic Review* 2, no. 2 (1950): 15.

NOTES

1. C.O. 260/62, SM Report, Grey to Stanley, Apr. 17, 1843.
2. C.O. 258/16, J.O. Drysdale, SM Report, Dec. 31, 1846-Dec. 31, 1848.
3. C.O. 258/16, SM Report, J. de Bernard to Drysdale, Jan. 15, 1847, p.162.
4. *The Voice of St Lucia*, (Aug. 15, 1885).
5. C.O. 258/16, SM Report, June 30, 1848, p.326.
6. Dispatches 1858-1863, p.57.
7. *The Port of Spain Gazette* (June 10, 1846).
8. C.O. 258/16, SM Report, J. de Bernard to Drysdale, Jan. 15, 1847.
9. C.O.253/73/3:1/1/1841.
10. C.O.253/76/2:10/5/1842.
11. *P.P.* 1847-1848, Reid to Grey, May 3, 1848.
12. C.O. 253/79, SM Report, Grey to Stanley, Sept. 11, 1843.
13. C.O. 258/16, SM Report, J. de Bernard to Drysdale, Jan. 15, 1847, p.162.
14. Dispatches from SOS Governor of the Windward Islands to St Lucia, 1848-1850.
15. C.O. 258/16, St Lucia Ordinances 1850, No. 2.
16. C.O.253/73/4:1/5/1841.
17. C.O. 253/83, SM Report, Grey to Stanley, April 1845.
18. C.O. 260/62, SM Report, Grey to Stanley, April 17, 1843.
19. C.O. 258/16, SM Report, Dec. 31, 1848.
20. *P.P.* 1847-1848 XLVI (749), Reid to Grey, May 3, 1848.
21. Blue Book of 1880-1889, pp.116, 176.
22. Caribbean Commission Crop Inquiry Series, No. 6 (London, H.M.S.O., 1949), p.175.
23. Blue Book of 1895, p.19.
24. Blue Book of 1889, p.8; Blue Book of 1892, p.7.
25. Blue Book of 1883, pp.118-19.
26. C.O.253/74:17/9/1841.
27. Royal Commission Report 1897, Vol. 8, p.71, Irish University Series, 1971.

28. Intercolonial Dispatches 1851-59 (London, H.M.S.O., 1864).
29. Dispatch No. 64 of 1864 (London, H.M.S.O.).
30. C.O. 321/191, Sendall to Holland, March 16, 1887.
31. Blue Book of 1889, p.10.
32. *The Voice of St Lucia* (1888).
33. C.O.253/128/1:25/5/1861.
34. C.O.253/74/5:17/9/1841.
35. Royal Commission Report 1897, Vol. 8, p.81, Irish University Press Series, 1971.
36. C.O. 253/82, SM Report, Grey to Stanley, May 7, 1844.
37. *Journal of Caribbean History*, Vol. 4 (May 1972), pp.46-47.
38. Royal Commission Report 1897, Vol. 8, p.89, Irish University Press Series, 1971.
39. Dispatch No. 18, 1896.
40. *West Indian Bulletin*, Vol. 1, p.16 (1889); Royal Commission Report 1897 (London, H.M.S.O.), Vol. 7.
41. Royal Commission Report 1897, Vol. 7, p.89, Irish University Press Series, 1971.
42. Interview with a landlord in Dennery, May 1989.
43. Blue Book of 1889, p.11.
44. Blue Book of 1921, p.25.
45. Land Settlement Scheme for St Lucia (London, H.M.S.O., 1958) p.6.
46. Blue Book of 1889, p.16.
47. Blue Book of 1889, p.4.
48. Blue Book of 1889, pp.17,18.
49. Royal Commission Report 1897, Vol. 8, p.89.
50. Dispatch No. 124 of 1907 (London, H.M.S.O.).
51. Royal Commission Report 1897, Vol. 8, p.71, Irish University Press Series, 1971.
52. Blue Book of 1897, p.17.
53. Interview with a sharecropper in Dennery, May 1989.
54. *Report on the Sugar Industry of the West Indies and British Guiana*, West Indies Sugar Commission 1929-1930 (London, H.M.S.O.), p.88.
55. Dispatch No. 106 of 1908 (London, H.M.S.O.), p.411.
56. Dispatch No. 10 of 1909 (London, H.M.S.O.).
57. Blue Book of 1929.
58. "St Lucia at Independence", *The Voice of St Lucia* (Feb. 22, 1979), p.24.
59. Grenada Council Papers, Legislative Council, 1931, p.2.
60. *The Voice of St Lucia* (May 3, 1958).
61. *Caribbean Economic Review* 2, no. 2 (1950), p.15.
62. Colonial Reports, 1951 and 1952, p.27.
63. *Report of the Commission Appointed by His Excellency The Governor to Enquire into the Stoppage of Work at the Sugar Factories in St Lucia, in March 1952 and into the Adequacy of the Existing Wage-Fixing Machinery in the Colony* (London, H.M.S.O., 1952).
64. *Agriculture in the West Indies* (London, H.M.S.O., 1942), pp.178-79.
65. Blue Book of 1954, p.56.
66. Commission Report 1952, p.14.

67. St Lucia Colonial Office Report, 1951.
68. Commission Report 1952, p.14.
69. Colonial Office Report, 1955-1956.
70. Land Settlement Scheme for St Lucia, 1958, p.10.
71. Interview with a landlord in Dennery, May 1989.
72. C.O. 253/83/6:12/1/1845.
73. *British West Indies Sugar Association Report* (London, W. Heffer and Sons, Ltd, 1954), pp.94, 99.
74. Interview with a sharecropper in Dennery, May 1989.
75. Interview with District Tax Collector.
76. Sugar Association Report, 1954, pp.94, 99.
77. Colonial Office Report, 1953, 1954.
78. Colonial Office Report, 1955-1956, p.31.
79. *The Voice of St Lucia* (Feb. 21, 1957), p.18.
80. Colonial Office Report, 1957, p.28.
81. Colonial Office Report, 1963, p.32.
82. Blue Book of 1937.
83. *St Lucia Gazette* (1953-1958).

BIBLIOGRAPHY

Primary Sources

British Colonial Office Records

C.O. 253/76/2:10/5/1842.

C.O. 253/53/3:1/3/1841.

C.O. 253/73/4:1/5/1841.

C.O. 253/99/1:15/6/1849.

C.O. 253/74/5:17/9/1841.

C.O. 253/238/1:25/5/1861.

C.O. 253/79/3:7/8/1843.

C.O. 253/83/6:12/1/1845.

C.O. 253/79 Grey to Stanley, Sept. 11, 1843.

C.O. 290/4 Tobago SM Report, Dec. 1853.

C.O. 258/2 *St Lucia Palladium* Dec. 3, 1853.

C.O. 258/16 SM Report, J. de Bernard to Drysdale, Jan. 15, 1847.

C.O. 258/16 J.O. Drysdale, SM Report, Dec. 31, 1846 -- Dec. 31, 1848.

C.O. 253/78 Grey to Stanley, April 24, 1843.

C.O. 101/101 Colebroke to Earl Grey, Oct. 13, 1849.

C.O. 253/103 Colebroke to Earl Grey, July 13, 1850.

C.O. 253/17, Lt. Gov. Darling, Despatch No. 115, Sept. 3, 1850.

National Archival Records

Colonial Reports -- Arrival No. 1929. In *Annual Report on the Social and Economic Progress of the People of St Lucia, 1938*. London, Her Majesty's Stationery Office (1940).

Dispatches from SOS to Lieutenant Governor of St Lucia, 1844-1846.

Grenada Council Papers, Legislative Council, 1931.

Intercolonial Dispatches, 1851-1914.

Lieutenant Governor Drysdale's Dispatches, Tobago, 1856-1862.

Report of the Commission Appointed by His Excellency The Governor to Enquire into the Stoppage of Work at the Sugar Factories in St Lucia, in March 1952 and into the Adequacy of the Existing Wage-Fixing Machinery in the Colony.

Report on the Census of the Colony of St Lucia 1921.

Report on the Sugar Industry of The West Indies and British Guiana, West Indies Sugar Commission 1929-30.

Royal Commission Report 1897 Vols. 6-8. Irish University Press Series, 1971.

St Lucia Blue Book of Statistics, 1893-1956.

Statistics for Canes Supplied to Each Factory by Company Estates, Contributors and Metayers for 1951 Crop.

St Lucia Colonial Office Reports, 1929-1963.

St Lucia Gazette, 1867-1958.

St Lucia Palladium, 1838-1839.

The Voice of St Lucia, 1885-1957.

West India Royal Commission Report 1945.

Books and Articles

Abalkin, L. et al. 1983. *Political Economy*. Moscow: Progress Publishers.

Acosta, Y., & J. Casmir. 1985. "Social orgins of the counter-plantation system in St Lucia". In *Rural Development in the Caribbean*, edited by P.I. Gomes. Kingston, Jamaica: Heinemann Educational Books.

Adelman, I., & C.T. Morris. 1975. *Economic Growth and Social Equity in Developing Countries*. Stanford: Stanford University Press.

Albert, M., & R. Hahnel. 1978. *Unorthodox Marxism*. Boston: South End Press.

Augier, F.R., et al. 1960. *The Making of the West Indies*. Kingston, Jamaica: Longman Caribbean.

Bagehi, A. 1976. "Cropsharing tenancy and neo-classical economics". *Economic and Political Weekly* (Jan. 17).

Bardlan, P., & T. Srinivisan. 1971. "Cropsharing tenancy in agriculture: a theoretical and empirical analysis". *American Economic Review* (March).

Beckford, G. 1975. "Caribbean rural economy". In *Caribbean Economy*, edited by G. Beckford. Mona, Jamaica: University of the West Indies, Institute of Social and Economic Research.

Beckford, G. 1983. *Persistent Poverty*. London: Zed Books.

Bergad, L.W. 1983. *Coffee and the Growth of Agrarian Capitalism in Nineteenth Century Puerto Rico*. Princeton: Princeton University Press.

Bernstein, H. 1987. "Capitalism and petty commodity production". *The Journal of Social Studies* 36 (April).

Bharadwaj, K., & P.K. Das. 1975. "Tenurial conditions and mode of exploitation: a study of some villages in Orissa". *Economic and Political Weekly* (February).

Brass, T. 1986. "Unfree labour and capitalist restructuring in the agrarian sector: Peru and India". *Journal of Peasant Studies* 14, no. 1.

Breen, H.H. 1844. *St Lucia: Historical, Statistical and Descriptive.* London: Frank Cass.

*British West Indies Sugar Association (Inc)***.** 1954. London: W. Heffer and Sons.

Brook, C. 1976. *St Lucia.* London: Macmillan.

Byres, T.J. 1983. "Historical perspectives on sharecropping". *Journal of Peasant Studies* 10, nos. 2 & 3.

Caribbean Economic Review, 2, no. 2 (1950).

Carnegie, C.V. 1981. *The Rural Population of Saint Lucia: a Preliminary Analysis.* Department of Regional Development, Organization of American States, Barbados.

Cenac, W. 1988. *Coutume de Paris to 1988: the Evolution of Land Law in St Lucia.* Castries, St Lucia: Voice Press.

Cerrea, M.M. 1985. *Putting People First: Sociological Variables in Rural Development.* London: Oxford University Press.

Cheung, S.N. 1969. *Theory of Share Tenancy.* Chicago: University of Chicago Press.

Cross, M., & G. Heuman. 1988. *Labour in the Caribbean: from Emancipation to Independence.* London: Macmillan.

Currie, K., & L. Ray. 1985. "Class formation within the peasantry: recent theoretical development in the analysis of the peasantry with specific reference to the East African debate". *Sociology* 19, no. 4.

Davies, O. 1987. *The State in the Caribbean Society.* Mona, Jamaica: Insitute of Social and Economic Research.

Davy, J. 1854. *The West Indies Before and Since Emancipation.* London: Frank Cass.

De Guiran, L. 1986. "Elements of the agrarian history of the island of St Lucia: from the Caribs to the morrow of independence". Mimeographed, National Research and Development Foundation, St Lucia.

de Janvry, A. 1985. *The Agrarian Question and Reformism in Latin America.* Baltimore: Johns Hopkins University Press.

Dookhan, I. 1977. *A Post-Emancipation History of the West Indies.* London: Collins.

Emmanuel, P. 1981. "Socio-political aspects of the tenurial system in the Eastern Caribbean". Paper presented at Workshop on Land Tenure Issues in the Eastern Caribbean, 26-29 May, St Lucia.

Firth, R., & B.S. Yamey. 1969. *Capital, Saving and Credit in Peasant Societies.* London: George Allen and Unwin.

Frank, A.G. 1979. *Dependent Accumulation and Underdevelopment.* New York: Monthly Review Press.

Frucht, R.A. 1968. "Caribbean social type: neither peasant nor proletarian". *Social and Economic Studies* 17, no. 3.

Gaitskell, A. 1959. *Gezira: a Story of Development in Sudan*. London: Faber and Faber.

Galli, R.E. 1981. *The Political Economy of Rural Development, Peasants, International Capital and the State*. Albany: State University of New York Press.

Gatchet, C. 1975. *A History of the Roman Catholic Church in St Lucia*. Port of Spain, Trinidad: Key Caribbean Publications.

Gomes, P.I. 1985. *Rural Development in the Caribbean*. Kingston, Jamaica: Heinemann.

Green, W.A. 1981. *British Slave Emancipation: the Sugar Colonies and the Great Experiment 1830-1865*. Oxford: Oxford University Press.

Guillebaud, C.W. 1930. "West Indian sugar crises 1800-1930". *Journal of the Royal Society* 21, no. 9.

Hall, D.G. 1959. *Free Jamaica 1838-1865: an Economic History*. New Haven: Yale University Press.

Hay, L.G. 1884. *A Handbook of the Colony of Tobago*. Tobago: Government Printing Office.

Helminiak, T.W. 1972. *The Sugar-Banana Shift on St Lucia West Indies: Bilharzia and Malaria Disease Causal Linkages*. London: University Micro Films.

Hilton, R. 1976. *The Transition From Feudalism to Capitalism*. London: New Left Books.

Howard, J. 1984. "The share system in the Bahamas in the nineteenth and early twentieth centuries". *Slavery & Abolition* 5, no. 2.

"Land settlement scheme for Saint Lucia based on a survey of the agricultural and social conditions of the island, 1958". London: H.M.S.O.

Latin American Bureau. 1987. *Green Gold: Bananas and Dependency in the Eastern Caribbean*. London, England: Latin American Bureau Ltd.

Lemchard, L., & K. Legg. 1972. "Political clientelism and development: a preliminary analysis". *Comparative Politics* 4, no. 2 (January).

Lenin, V.I. 1964. *The Development of Capitalism in Russia*. Moscow: Progress Publishers.

Lenin, V.I. 1971. *A Great Beginning*. Moscow: Progress Publishers.

Lenin, V.I. 1980. *Selected Works*. Vol. 3. New York: International Publishers.

Lewis, W.A. 1936. "The evolution of a peasantry in the British West Indies". Colonial Office Pamphlet 656. London: H.M.S.O.

Llambi, L. 1988. "Small modern farmers: neither peasants nor fully fledged capitalists? A theoretical and historical discussion". *Journal of Peasant Studies* 15, no. 3.

Lobdell, R.A. 1988. "British officials and the West Indian peasantry 1842-1938". In *Labour in the Caribbean: From Emancipation to Independence*, edited by M. Cross and G. Heuman. London: Macmillan.

Louis, M. 1981. "An equal right to the soil: the emergence of a peasantry in St Lucia 1838-1900". PhD diss., Johns Hopkins University, Baltimore.

Marshall, A. 1961. *Principles of Economics*. London: Macmillan.

Marshall, W.K. 1963. "The social and economic development of the Windward Islands 1838-1865". PhD diss., Cambridge University, England.
Marx, K. 1967. *Capital.* Vol. 3. New York: International Publishers.
Marx, K. 1968. *Theories of Surplus Value.* Pt.1. Moscow: Progress Publishers.
Marx, K. 1973. *Grundrisse.* New York: Vintage Books.
Marx, K. 1980. *Pre-capitalist Economic Formations.* New York: International Publishers.
Marx, K. 1986. *Capital.* Vol. 1. Moscow: Progress Publishers.
Merton, R.K. 1949. "The role of applied social science in the formation of policy: a research memorandum". *Philosophy of Science* 16, no. 3.
Migdad, J.S. 1977. *Peasants, Politics and Revolutions: Pressures Towards Political and Social Change in the Third World.* Princeton: Princeton University Press.
Mill, J.S. 1848. *Principles of Political Economy.* London: Green and Co.
Mintz, S. 1961. "The question of Caribbean peasantries: a comment". *Caribbean Studies* 1, no. 3.
Mintz, S. 1974. *Caribbean Transformations.* Baltimore: Johns Hopkins University Press.
Munroe, T. 1980. "A B C of scientific socialism: a series of lectures". Mimeographed.
Nomsen, J.D. 1972. "Land tenure as a barrier to agricultural innovation: the case of St Lucia". Proceedings of the seventh West Indian Agricultural Economics Conference, Grenada.
Otsuka, K., and Y. Hyami. 1988. "Theories of share tenancy: a critical survey". *Economic Development and Cultural Change* 37, no. 1.
Pearce, R. 1983. "Sharecropping: towards a Marxist view". *Journal of Peasant Studies* 10, nos. 2 & 3.
Post, K. 1978. *Arise Ye Starvelings: the Jamaican Labour Rebellion of 1938 and its Aftermath.* The Hague, Netherlands: Martinus Nijhoff.
"Progress report on the agricultural development programme for St Lucia for the period ending 30 June, 1954". London: H.M.S.O.
Reid, J.D. 1975. "Sharecropping in history and theory". *Agricultural History* 49, no. 2.
Reid, J.D. 1977. "The theory of share tenancy revisited--again". *Journal of Political Economy* 85, no. 2.
Riviere, B. 1989. "Notes on the 'peasantry' in Dominica and questions concerning organization and mobilization". Mimeographed.
Romalis, R. 1968. "The rural entrepreneur in St Lucia". Occasional Paper, Series 5. Centre for Development Area Studies, Montreal: McGill University.
Sahr, W. 1987. *Agroecosystems of Small Farmers and their Socioeconomic Impact: a Case Study of Tete Chemin, St Lucia.* Fort Lauderdale: Verlag Breistenbach.
Selsam, H., and H. Martel. 1977. *Reader in Marxist Philosophy.* New York: International Publishers.
Shafer, R.J. 1971. *A Guide to Historical Method.* Illinois: Dorsey Press.

Sismonde de Sismondi, J.C.L. 1966. *Political Economy*. New York: Augustus M. Kelly.

Smith, A. 1976. *An Enquiry into the Nature and Causes of the Wealth of Nations*. Chicago: University of Chicago Press.

Stavenhagen, R. 1975. *Social Classes in Agrarian Societies*. New York: Anchor Press.

St Lucia Economic and Social Review. 1983-1988. Castries, St Lucia: Ministry of Finance.

St Lucia Human Settlements. 1988. Washington: Department of Regional Development, Organization of American States.

"St Lucia. 1957. Revised development sketch plan 1946-1956". Mimeographed.

Sweezy, P.M. 1970. *The Theory of Capitalist Development: Principles of Marxian Political Economy*. New York: Monthly Review Press.

"The sugar industry in the Caribbean". 1947. Caribbean Commission Crop Inquiry, Series No. 6. London: H.M.S.O.

Thomas, C.Y. 1984. *Plantations, Peasants and State*. University of the West Indies, Mona: Institute of Social and Economic Research.

Thomas, C.Y. 1988. *The Poor and the Powerless: Economic Policy and Change in the Caribbean*. New York: Monthly Review Press.

West Indian Bulletin (Journal of the Imperial Agricultural Department for the West Indies) 1, no. 1 (1899).

Williams, E. 1964. *Capitalism and Slavery*. London: André Deutsch.